D0713218

Physics Literature

A Reference Manual

by Robert H. Whitford, M.E., M.S., Ed.D.

SECOND EDITION

The Scarecrow Press ☆ Metuchen, N. J. ☆ 1968

Library of Congress Card No. 68-12636

To

My wife, LILIAN

PREFACE

This is a survey of physics literature at the college level. It describes the many types and forms available, selects a representative working collection, and outlines efficient library methods. Arrangement is by most usual lines of inquiry, termed "approaches," e.g., the historical approach when items pertaining to history are sought. Background materials have been interspersed for greater interest and information.

In his efforts to produce a helpful guide, the writer has drawn upon long experience in an active science-technology library, with an interlude of college physics teaching.

In this revised edition the literature has been resurveyed. Necessary modifications, additions and deletions have been made, while preserving bibliographical and textual passages still applicable.

Robert H. Whitford
Assoc. Professor and Librarian
Engineering and Science Library
The City College, New York

TABLE OF CONTENTS

CHAPTER I

INTRODUCTION

Physics is the subject of concentration of this guide; material from related fields is cited only when necessary for background purposes. Most books listed are at the college level; lucid comprehensive presentations readily absorbed by college students usually take precedence over advanced mathematical monographs intelligible only to specialists. Of course, a specialist seeking data outside his own familiar field will find these helpful.

Scope.

A guide to special subject literature should give representative sources and indicate how to proceed further, for indiscriminate listing of all possible references would be a disservice to users seeking orientation and guidance. Accordingly, this guide comprises material selected according to the following criteria:

(1) Is it a useful bibliographical tool? (For example, a periodical indexing medium.)

(2) Is it a comprehensive reference work? (These are designed to yield needed data rather than for reading through.)

(3) Does it sketch a particular aspect well? (For example—biographical.)

(4) Does it fill a major subject gap? (E.g., ultrasonics.)

(5) Is it a recent publication? (These usually supersede and review previous work.)

(6) Might it be termed a "classic?" (A time-honored book possessing lasting utility, as well as historical interest.)

(7) Has it reached the multi-edition stage? (Thereby denoting widespread acceptance through merit.)

(8) Written by an authority in his field? (E.g., Bridgman for high pressure; Aston for mass spectra.)

(9) Is it otherwise a noteworthy publication?

Besides indicating informational sources and outlining special techniques, this guide sketches physics literature in general, so as to call attention to the different kinds of printed materials available but often overlooked.

Omission of a particular book does not imply lack of merit, as this is not a roster of "best books." Instead, the books cited are those most useful for present purposes, *viz.*, to furnish helpful guidance and a basic collection of sources. Sometimes equally good titles have been omitted for quantitative reasons, as in the field of mechanics which is so profusely covered by textbooks differing in arrangement, style, and scope. No compiler being infallible, some important books have possibly been overlooked. However, these inadvertent omissions have been minimized by employing parallel methods of compilation. Library searches were supplemented by gleanings from publishers' catalogs, book reviews, bibliographies, etc. Moreover, the writer's many years of active service in a technical library furnished first-hand subjective knowledge of most useful references.

As Jones states, "Any guide must be selective, partial, and fairly quickly obsolescent."[1] Hence the user cannot hope to find every worthwhile book listed among the pages of this guide, especially if publication dates prevent inclusion.

Arrangement.

The entire guide is arranged according to a multi-approach pattern, based on the various aspects of physics that an information seeker may have in mind. For example, he may wish biographical, experimental, or terminological material, or he may be concerned with educational implications. Such phases are termed approaches, because they represent usual lines of inquiry. In addition, there is the topical approach, pursued whenever items on a physics subject *per se* are sought.

In each chapter explanatory text and book citations have been organized into a connected account so as to lead the reader from one group of references to the next, preserving continuity and facilitating comparison. Relevant background material has been interspersed throughout to increase general interest and usefulness. Descriptive notes are given for individual books whenever special features, not indicated by their grouping or titles, merit attention. Overviews and summaries round out and coordinate the whole.

Research aspects.

Whitney has accorded recognition to library and documentary research as follows:

---

1. E. L. Jones, "Searching the Literature of Science." *Journal of Scientific Instruments,* 17:253-257, November 1940.

Descriptive research may be in terms of surveys and critical analyses of available data in printed form. This is informational analysis, or library research as it is called at times. It constitutes one technique of historical research, as history deals with records of the past. Obviously, making bibliographical lists is not research; but a critical evaluation of a unitary group of material with interpretation in terms of comparison and generalization may employ reflective thinking.[2]

This guide represents a borderline contribution between the fields of physics and library science. Spratt boasts of this interlinkage of subject specialization and librarianship in following vein:

> As a scientific librarian, it would be shameful of me to claim that our qualifications were above those for other branches of librarianship; but it must be admitted that with Scientific literature, as opposed to what the Germans would call "Schöne Literatur," the subject-matter aspect is more important than the author. Now the compilation of an alphabetical author-index is more or less routine work, but the precise subject-matter classification of Scientific literature requires one to be specialist as well as librarian; in short, a *super*-librarian.[3]

Finally, the chapter on the Educational Approach is of particular significance to educators engaged in science teaching.

---

2. F. L. Whitney, *The Elements of Research,* p. 178. (Third Edition.) New York: Prentice-Hall, Inc., 1950.
3. H. P. Spratt, *Libraries for Scientific Research . . .* , p. 10-11. London: Grafton and Company, 1936.

CHAPTER II

BIBLIOGRAPHICAL APPROACH

Serials, Books, and Library Usage

The "bibliographical approach" is here taken to signify a broad, general approach to physics literature in periodical or book form, rather than preoccupation with bibliographical *lists per se*. This alternative is sanctioned by one Funk and Wagnalls definition of bibliography as works collectively bearing on a particular subject, whether listed or not.

1—Serial Publications

Periodicals are publications that appear in a continuing series of issues, usually at regular intervals, under the same distinctive title. Serials are likely to have discrete characteristics with respect to content, title and publication date. Whenever units of a serial or pseudo-serial set are shelved together in sequence within a library, they may be requested and located most conveniently from the series note on the catalog card (in parentheses after pagination, sometimes conspicuously underlined in red). For example, see the entry for H. C. Bolton's *Catalogue* referring to *Smithsonian Miscellaneous Collections Vol. 40.* (Excepted are the scattered books of commercial series such as *Methuen's Monographs.*)

General.

One turns to periodicals for the latest work in a given field, and for material on topics too small to warrant book treatment. In general, books are behind current practice because of compilation and publication delays. As Spratt declares:

> To collect the material for a comprehensive text-book, to hammer it into a presentable manuscript, is tedious work for the conscientious author, and cannot be hurried. It follows that when his book has at last been printed and published, it is already out-of-date for the scientific research worker who wants to know what was done last week in his particular field. For such readers, continual reference must be made to the most recent articles in current periodicals.[1]

---

1. H. P. Spratt, *op. cit.*, p. 10.

For general introduction see:

a. American Institute of Physics. *The Periodical Literature of Physics: A Handbook for Graduate Students,* by Robert E. Maizell and Frieda Siegel. New York: The Institute, 1961. 15 pp.

b. Brown, Charles H. *Scientific Serials.* Chicago: Association of College and Research Libraries, 1956. 189 pp. (ACRL Monograph No. 16.)

c. Kronick, David A. *A History of Scientific and Technical Periodicals: The Origins and Development of the Scientific and Technological Press, 1665-1790.* New York: The Scarecrow Press, 1962. 274 pp.

There are subject guides to current journals:

a. *Ulrich's International Periodicals Directory.* Vol. 1: *Scientific, Technical and Medical.* (Twelfth Edition.) Edited by Marietta Chicorel. New York: R. R. Bowker Company, 1967. 540 pp.
"A classified guide to a selected list of current periodicals, foreign and domestic." Arts, humanities, social sciences and business are covered in the second volume, and both are updated by the annual supplements, alternately. There is a companion directory, *Irregular Serials and Annuals.*

b. *Standard Periodical Directory, 1967.* (Second Edition.) United States and Canadian Periodicals. New York: Oxbridge Publishing Co., 1966. 1019 pp.
The coverage is comprehensive.

Lists of American and foreign scientific serials have been issued by the Library of Congress.

To ascertain which library possesses files of particular journals, as well as for checking publication details, one consults union lists such as the following general compilation:

Union List of Serials in Libraries of the United States and Canada, edited by Edna B. Titus. (Third Edition.) New York: The H. W. Wilson Company, 1965. 5 vols.
Library holdings are keyed to symbols listed in the preliminary pages of volumes; e.g., NNE designates the Engineering Societies Library, New York.

The foregoing is supplemented by the monthly *New Serial Titles,* cumulating annually and for multiyear periods, now 1950-1960 and 1961-1965.

Holdings of smaller technical libraries are surveyed by:

Special Libraries Association. *Union List of Technical Periodicals in Two Hundred Libraries of the Science-Technology Group of the Special*

Libraries Association, compiled by E. G. Bowerman. (Third Edition.) New York: The Association, 1947. 285 pp.

A comprehensive British union list is:

World List of Scientific Periodicals Published in the Years 1900-1960, edited by Peter Brown and George B. Stratton. (Fourth Edition.) London: Butterworths Scientific Publications, 1963-1965. 3 vols. This final edition includes 60,000 titles, and its continued updating will be by *British Union-Catalogue of Periodicals* supplements. It furnishes a means for identifying journals, and indicates British library holdings. Title abbreviations conform with the new British Standard.[2]

Two earlier lists of science journals, covering practically identical material, which are useful for data on publications of the period, are:

a. Bolton, Henry C. *A Catalogue of Scientific and Technical Periodicals 1665-1895, together with Chronological Tables and a Library Checklist.* (Second Edition.) Washington, D. C.: Smithsonian Institution, 1897. 1247 pp. (Smithsonian Miscellaneous Collections Vol. 40.)
A subject index to the alphabetical listing enables one to select periodicals on "Physics" and other topics.

b. Scudder, Samuel H. *Catalogue of Scientific Serials of All Countries including the Transactions of Learned Societies in the Natural, Physical and Mathematical Sciences 1633-1876.* Cambridge, Mass.: Library of Harvard University, 1879. 358 pp.
This is geographically arranged by country and town.

For current periodicals, there are convenient lists (indicating coverage), issued by the indexing and abstracting media such as *Chemical Abstracts, Physics Abstracts*, etc.

Less recent is a *List of Periodicals of Physics Interest* [3] that indicates for each journal whether it is indexed in any of the "three abstracting services most frequently consulted by physicists . . . *Physics Abstracts, Chemical Abstracts*, and *Nuclear Science Abstracts*," and the proportion of its contents devoted to physics.

For useful data on United Kingdom scientific journals, see:

Royal Society of London. *A List of British Scientific Publications Reporting Original Work or Critical Reviews.* London: The Society, 1950. 95 pp.

---

2. See the *Chemical Abstracts* standard list of title abbreviations following USA Standard Z39.5-1963, adopted by the American Institute of Physics.
3. Compiled by Robert S. Bray, and available from U. S. Dept. of Commerce by number: PB 110082.

Periodical titles that include a society name are not inverted in AIP and ACS lists, *i.e., Journal of the Optical Society of America* is filed under "J." However, this title would appear under "O" according to ALA cataloging rules which place the society name first, *viz., Optical Society of America. Journal.* There will still be no inversion in *Chemical Abstracts'* new "Comprehensive List," but entries will include reference to the ALA forms.

Indexes and abstracts.

Annual indexes are appended to most of the bound volumes on library shelves, and publishers often issue combined indexes for several years output. These are useful. Noteworthy indexes are listed in:

a. Haskell, Daniel C. *A Check List of Cumulative Indexes to Individual Periodicals in the New York Public Library.* New York: The New York Public Library, 1942. 370 pp.

b. Ireland, Norma O. *An Index to Indexes; A Subject Bibliography of Published Indexes.* Boston: F. W. Faxon Company, 1942. 107 pp. (Unpublished indexes appear in her *Local Indexes . . .* Same publisher, 1947. 221 pp.)

Science indexing and abstracting media are listed in the following:

a. *Index Bibliographicus.* (Fourth Edition.) Vol. 1: *Science and Technology.* The Hague: Fédération International de Documentation, 1959. 118 pp. Convenient classified arrangement.

b. National Federation of Science Abstracting and Indexing Services. *A Guide to the World's Abstracting and Indexing Services in Science and Technology.* Washington, D. C.: The Federation, 1963. 183 pp.

This L.C. compilation updates the preceding item as well as another L.C. list for U. S. media (1960), all of which are later than Gray and Bray's convenient physics selection.[4]

As the foregoing lists cover indexing in special areas very comprehensively, only some major indexing and abstracting services need be described:

Science Abstracts: For the physicist, a good source of brief summaries (*i.e.,* abstracts) of articles is *Physics Abstracts,* in combination with *Electrical and Electronics Abstracts* and *Control Abstracts* constituting *Science Abstracts* (Sections A, B and C, respectively). This

---

4. D. E. Gray and R. S. Bray, "Abstracting and Indexing Services of Physics Interest." *American Journal of Physics,* 18: 274-299, May 1950. Additions and corrections: *Ibid,* 18: 578-579, December 1950.

tool has been published since 1898 under the auspices of British and American physical and electrical-engineering societies. Its present enlarged scope is indicated by the acronym INSPEC, standing for Information Service in Physics, Electrotechnology and Control. There are now semi-annual author-subject indexes and some multi-year indexes for this monthly medium. *Current Papers in Physics* began publication semi-monthly in 1966 as a complementary current-awareness service.

Applied Science and Technology Index: Volumes 1-45 (1913-1957) bore the title *Industrial Arts Index,* which somewhat obscured their high physics content. This convenient index is in alphabetical subject arrangement, without abstracts, and possesses the characteristic cumulative feature of H. W. Wilson Company indexes. Thus, at intervals throughout the year, groups of prior issues will be gathered into one alphabet for convenient reference, with a final cumulation at year's end. An article may receive multiple listing under several relevant subject headings, but there is no author approach. This index analyzes fully the contents of over 200 periodicals (listed in the prefatory pages of cumulative volumes).

Engineering Index: Formerly the annual volumes sufficed for those who could not accommodate the selective card-index service, but since 1962 there have been general monthly issues, and special *Electrical/Electronics* and *Plastics* sections since 1965. Arrangement is now alphabetical by major subjects and subheadings, linked by cross-references. (Before 1919, it was by classified subject.) Annual author indexes for the volumes since 1928 make finding articles by a particular individual easy. *Engineering Index* has been published since 1884, and selectively indexes the content of about 1500 journals which are listed on prefatory pages. Each article receives a listing under but one subject heading, accompanied by an abstract. Many of the subjects, particularly electrical applications, are of interest to physicists as well as to engineers. The British counterpart is *British Technology Index,* begun 1962.

Science Citation Index, published by the Institute for Scientific Information (Philadelphia) since 1961, is a new technique which Park describes as follows:

> To use it, one thinks of an author, or better still, a particular paper likely to be quoted in current literature on the subject in which one is interested. If anyone has referred to the author in the period covered, he will be found in the *Citation Index.* Below his name is listed in chronological order each cited work, together with the references to every

paper that refers to it. The idea is that, by looking up citations of these papers in turn, one can fill in a bibliography. A great merit of this process is that one will find unsuspected ramifications where an idea from one field of science has quietly been adopted in another. The main difficulty is that quotation of other authors is, at best, a capricious business; and in physics, after the first few years, the most seminal works are generally not cited at all because it is assumed that people are familiar with them.[5]

Applied Mechanics Reviews: This critical review medium indexes selected articles from five hundred magazines, etc., from all over the world, and began in 1948. It furnishes abstracts of important papers on theoretical and experimental aspects of fluid mechanics, solid mechanics, heat, gas dynamics, etc. Classified subject grouping is used, with annual author and subject indexes. The publisher is the American Society of Mechanical Engineers. WADEX is a supplementary word and author index available from the U. S. Government Printing Office.

Nuclear Science Abstracts and *U. S. Government R & D Reports* are discussed under Government Publications.

Solid State Abstracts Journal. This comprehensive abstracting medium is now published quarterly. Before 1967 it was the monthly *Solid State Abstracts,* published since 1960 by Cambridge Communications Corporation (Mass.) in continuation of its *Semiconductor Electronics.* Its headings range throughout the conventional subdivisions of physics, as applied to the solid state. It is paralleled by two similar quarterlies, *Electronics Abstracts Journal* and *Information Processing Journal,* assuring world literature coverage, theoretical and applied.

Electronic Engineering Master Index is now published by Master Index Services, Inc., New York. Since 1945 several multi-year volumes have appeared, the first spanning 1925-1945, and the latest covering 1949-1964. These are convenient supplementary indexes for an important field of engineering which has physics ramifications.

Physikalische Berichte, herausgegeben vom Verband Deutscher Physikalischer Gesellschaften. Braunschweig: Friedrich Vieweg und Sohn, 1920-
This monthly index parallels the English-language media, and continues *Fortschritte der Physik.*

Chemical Abstracts: Published by the American Chemical Society since 1907, this is without doubt the most comprehensive abstracting medium that exists in any field. Chemists rely upon it for informa-

---

5. D. Park, in review. *American Journal of Physics,* 34: 994, October 1966.

tion gleaned from thousands of periodicals throughout the world, including items of physics interest. Current weekly issues have a broad classified arrangement, with author, keyword and patent indexes. Contents for previous years are rendered conveniently accessible by an impressive array of indexes, including decennial and more recently quinquennial indexes, which obviate consultation of annual indexes for the years spanned. Through 1966, small superior letters (formerly numbers) found in indexes merely indicated how far down on the page an abstract appeared. Beginning 1967, individual abstracts are numbered consecutively. Symbol "P" is used for patent entries. The latest list of periodicals abstracted as of 1967 is comprised of the 1961 list plus its five supplements covering 1962 through June 1967. (A computer-based *Comprehensive List* is scheduled for publication towards the end of 1968.) Note that the alphabetization is according to the *bold-faced portions* of the journal titles, these constituting standard abbreviations: [6]

> CA's authorized journal-name abbreviations have been used for over fifty years by the American Chemical Society. They have also been adopted for journals by the American Institute of Physics. They are used by many individual writers, editors, and publishers.

Instrument Abstracts: Compiled under this title since 1959 by the British Scientific Instrument Research Association, and as their *Bulletin* beginning 1946.

Bulletin Signalétique: Published since 1940 by Centre National de la Recherche Scientifique, Paris. (Earlier title was *Bulletin Analytique.*) Separate parts for areas of physics and other sciences are issued monthly, since 1961.

Geophysical Abstracts is a quarterly publication of the U. S. Geological Survey. Issues numbered 1-191 (1929-1962) appeared either in its *Bulletin* or in the U. S. Bureau of Mines *Information Circular* series. Reprint publishers offer these scattered issues in collected form for more convenient use.

Meteorological and Geoastrophysical Abstracts has been published monthly by the American Meteorological Society, Boston, since 1950. (The title was originally *Meteorological Abstracts.*)

Scientific and Technical Aerospace Reports (STAR) is the semimonthly abstracting journal of the U. S. National Aeronautics and Space Administration (NASA). It began in 1963, superseding *Tech-*

---

6. Beginning 1967 conforming with *USA Standard for Periodical Title Abbreviations,* Z39.5-1963.

nical Publications Announcements. Reports are covered world-wide. The voluminous indexes are accompanied by a guide to the STAR subject indexes.

International Aerospace Abstracts is published semi-monthly by the American Institute of Aeronautics and Astronautics, and started in 1961. It covers published literature in books and journals rather than reports, thus complementing STAR.

Bibliographic Index: Beginning in 1938, this H. W. Wilson publication now appears semi-annually, with either an annual or three-year cumulation each December. It is a subject index to current bibliographies "including those published separately as books and pamphlets, and those published as parts of books, pamphlets and periodical articles." Although general in scope, it includes items of physics interest.

Other general indexing services useful on occasion to physics readers are the H. W. Wilson Company's *Readers' Guide to Periodical Literature, Education Index,* etc. For translation listings, see Chapter VIII. For photography, etc., see Eastman Kodak's *Abstracts of Photographic Science and Engineering Literature,* formerly *Monthly Abstract Bulletin.*

Earlier material is spanned chronologically by the following trio:

a. Reuss, Jeremias D. *Repertorium Commentationum a Societatibus Litterariis Editarum.* Gottingae: Henricum Dietrich, 1801-1821. 16 vols.

Reprinted: N. Y., Burt Franklin, 1961. Of special interest is Vol. 4: Physica (1805). Indexing of society publications is by classified subject, with an author index. The language of the article (including English) is used in its entry. (*See also* Thomas Young's great bibliography for parallel coverage.)

b. Royal Society of London. *Catalogue of Scientific Papers, 1800-1900.* London: C. J. Clay and Sons, 1867-1902; Cambridge, England: At the University Press, 1914-1925. 19 vols.

Subject Index: Vol. 1, Mathematics; Vol. 2, Mechanics; Vol. 3 (in two parts), Physics. Cambridge, England: At the University Press, 1908-1914. 3 vols. in 4.

This combination forms a monumental author and subject approach to articles in periodicals and transactions of societies for the whole nineteenth century. Fortunately for the physicist, the index volumes for his major fields of interest have been the first to be published. These are arranged by classified subjects, according to the *International Catalogue of Scientific Literature* schedules (outlined at beginnings of volumes and indexed at ends). The articles are arranged chronologically under subject. The index volumes have complete data (except title in full), and lists of journals covered.

c. *International Catalogue of Scientific Literature, 1901-1914/1916.* Published for the International Council by the Royal Society of London. London: Harrison, 1902-1919. 14 annual issues, each in 17 vols. While published, this was a science bibliography of truly mammoth proportions. Each of the seventeen annual volumes was devoted to a science, e.g., A—Mathematics; B—Mechanics; C—Physics; etc., classified according to the listed schedules, and with author and subject approaches to book and periodical literature.

Reviews.

Unlike the indexing and abstracting media, which identify individual articles, review serials endeavor to synthesize important researches into a connected account of scientific progress. They represent further assistance to the physicist in his attempt to assimilate current material relevant to his subject field. Three general media are:

a. *Reviews of Modern Physics.* New York: American Institute of Physics, 1929 to date. This is a quarterly publication under sponsorship of the American Physical Society, and furnishes well-documented surveys of research in various areas.

b. *Reports on Progress in Physics.* London: Institute of Physics and the Physical Society, 1934 to date. These are very interesting bibliographical accounts of yearly advance along many lines, including educational. Some articles even cover the whole development of a subject.

c. *Ergebnisse der Exakten Naturwissenschaften.* This series of review monographs began in 1922. There is a cumulative index to vols. 11-27 in Vol. 27 (Berlin: Springer, 1953. 421 pp.) For vols. 1-10 consult Vol. 22.

With the ever-rising flood of book and periodical literature, it becomes increasingly difficult for scientists and engineers to keep abreast of new developments, especially in fields that border their own. The appearance of special reviewing series seemed to be the most promising form of assistance, even though these sometimes resemble journals rather than critical reviewing media designed to furnish overall orientation. The articles merely tend to be longer without reaching monographic proportions. Perhaps more unified treatment would help:

> This idea of focusing a given volume of these series variously entitled: Advances in, Progress in, Studies in, Reviews of, etc., on a particular

subject is an excellent one which deserves a much wider practice. What is particularly important in producing a successful focused review volume is the careful choice of an editor and his careful choice of authors with the concomitant selection of live topics in the forefront of research.[7]

So numerous have these series become that only some of the more familiar ones will be cited below. Whether the word "Advances" or "Progress" heads the title is indicated within parentheses, and the following abbreviations are used for publishers:

AP: Academic Press, New York.
AR: Annual Reviews, Inc., Stanford, California.
IP: Interscience Publishers, Division of JW.
JW: John Wiley and Sons, New York.
PP: Pergamon Press, Oxford, New York, etc.

Applied Mechanics (Adv) AP 1948-
Astronautical Sciences (Prog) IP 1962-
Astronautics and Aeronautics (Prog) AP 1960-
Astronomy and Astrophysics (Adv) AP 1962-
Atomic and Molecular Physics (Adv) AP 1965-
Biological and Medical Physics (Adv) AP 1948-
Biophysics and Molecular Biology (Prog) [8] PP 1950-
Chemical Physics (Adv) IP 1958-
Computers (Adv) AP 1960-
Cryogenics (Prog) AP 1959-
Dielectrics (Prog) JW 1959-61; AP 1962-
Electronics and Electron Physics (Adv) AP 1948-
Elem. Particle and Cosmic Ray Physics (Prog) IP 1952-
Geophysics (Adv) AP 1952-
High Pressure Research (Adv) AP 1966-
Low Temperature Physics (Prog) IP 1955-
Magnetic Resonance (Adv) AP 1965-
Materials Science [*Metal Physics*] (Prog) PP 1949-
Microwaves (Adv) AP 1966-
Nuclear Energy [12 series] (Prog) PP 1956-
Nuclear Physics (Prog) PP 1950-
Nuclear Science and Technology (Adv) AP 1962-
Nuclear Techniques and Instrumentation (Prog) IP 1965-
Optical and Electron Microscopy (Adv) AP 1966-
Optics (Prog) IP 1961-
Radio Research (Adv) AP 1964-
Semiconductors (Prog) JW 1956-
Solid Mechanics (Prog) IP 1960-
Space Science and Technology (Adv) AP 1959-

---

7. K. E. Shuler, in review. *Physics Today*, 19 (No. 9): 91, September 1966.
8. Title changed (1964) from *Progress in Biophysics and Biophysical Chemistry*, originated by AP.

Spectroscopy (Adv) IP 1959-
Theoretical Physics (Adv) AP 1965-

Some reviewing series do not have the "advances" designation in their main title, e.g.:

Annual Review of Astronomy and Astrophysics AR 1963-
Annual Review of Nuclear Science AR 1952-
Methods in Computational Physics AP 1963-
Physics of Thin Films AP 1963-
Solid State Physics AP 1955-
Studies in Statistical Mechanics IP 1962-

See also Applied Mechanics Reviews, and *Technical Progress Reviews.*

Advances in Physics is a separate quarterly supplement of the *Philosophical Magazine* (London) beginning with the January 1952 issue.

Readable and informative review articles are frequently featured in the *American Journal of Physics,* acquainting students and non-specialists with a subject (e.g., masers and lasers [9]) within relatively small compass. *See also* certain sources mentioned under Current Events and various treatises throughout the Topical Approach.

Societies.

Society publications represent scholarly contributions to research, and serve as valuable supplements to other books and journals. Issuing agencies are listed in various guides with descriptive notes on history, object, membership, activities and publications:

a. National Research Council.[10] *Scientific and Technical Societies of the United States and Canada.* (Seventh Edition.) Washington, D. C.: The Council, 1961. 497 pp. (Publication 900.)

b. *Scientific and Learned Societies of Great Britain: A Handbook Compiled from Official Sources.* (Sixty-First Edition.) London: George Allen and Unwin, Ltd., 1964. 222 pp.

Learned societies (as well as universities, colleges, technical schools, etc.) are listed world-wide in the German *Minerva,* and in:

The World of Learning 1966-67. (Seventeenth Edition.) London: Europa Publications, Ltd., 1967. 1578 pp.

---

9. B. A. Lengyel, "Evolution of Masers and Lasers." *American Journal of Physics,* 34: 903-913, October 1966.
10. See D. E. Gray, "The National Academy of Sciences and the National Research Council." *Physics Today,* 5 (No. 1): 20-23, January 1952.

Historical treatments include:

a. Ornstein, Martha. *The Role of Scientific Societies in the Seventeenth Century.* (Third Edition.) Chicago: University of Chicago Press, 1938. 308 pp.

b. Bates, Ralph S. *Scientific Societies in the United States.* (Third Edition.) Cambridge, Mass.: The M.I.T. Press, 1965. 326 pp.

Full-length histories of particular societies may be represented by:

Lyons, Sir Henry G. *The Royal Society, 1660-1940.* Cambridge, England: At the University Press, 1944. 354 pp.

The American Institute of Physics and member societies (American Physical Society, Optical Society of America, Acoustical Society of America, Society of Rheology, and American Association of Physics Teachers) are described at length in *Physics Today,* 4 (No. 10): 12-27, October 1951.

International organizations, such as the International Union of Pure and Applied Physics, are listed in:

United States. Library of Congress. General Reference and Bibliography Division. *International Scientific Organizations; A Guide to Their Library, Documentation, and Information Services,* prepared under the direction of Kathrine O. Murra. Washington, D. C.: Library of Congress, 1962. 794 pp.

See also Brauer's *Forschungsinstitute, European Research Index,* etc.

Early society publications may be traced by means of the Reuss' *Repertorium,* Royal Society *Catalogue,* and *International Catalogue,* supplemented by regional bibliographies such as:

a. Müller, Johannes. *Die Wissenschaftlichen Vereine und Gesellschaften Deutschlands im Neunzehnten Jahrhundert; Bibliographie ihrer Veroffentlichungen seit ihrer Begründung bis auf die Gegenwart.* Berlin: Behrend and Company, 1883-1917. 2 vols. in 3.

b. Deniker, Joseph, and Descharmes, René. *Bibliographie des Travaux Scientifiques (Sciences Mathématiques, Physiques et Naturelles) publiés par les Sociétés Savantes de la France.* Paris: Imprimerie Nationale, 1895-1922. Vol. 1-Vol. 2, pt. 1, A—Sarthe.

Conferences.

There has been a tremendous increase in the number of published conference and symposium reports. Some are poorly-produced unedited accounts lacking indexes; others appear too late and at too high a cost considering their passing value. Among them, however, are useful comparative summaries of accomplishments and trends.

Current indexing media include:

a. *Directory of Published Proceedings,* issued ten times yearly, with annual cumulations purchasable separately, by InterDok of White Plains, N. Y. This indicates availability of actually-published proceedings of conferences, symposia, meetings and congresses for science and technology, internationally.

b. *World Meetings,* revised and cumulated quarterly, with five indexes (date, keyword, location, deadline, and sponsor). Separately published parts cover future medical, scientific and technical meetings held in the United States and Canada, and those held elsewhere, respectively. The former section began as *TMIS Technical Meetings Index.* Information on publications and exhibits is included.

The Library of Congress issues a quarterly *World List of Future International Meetings,* revised monthly, of which Part 1 includes science and technology. The Special Libraries Association publishes future *Scientific Meetings,* alphabetically and chronologically arranged, three times yearly.

An earlier compilation similar in function to the *Union List of Serials* was:

International Congresses and Conferences 1840-1937; A Union List of Their Publications Available in Libraries of the United States and Canada, edited by Winifred Gregory. New York: The H. W. Wilson Company, 1938. 229 pp.

Substantial volumes have been published for various international physics conferences; for example, the following titles:

a. *Rapports présentés au Congrès International de Physique réuni à Paris en 1900 sous les auspices de la Société Française de Physique.* Paris: Gauthier-Villars, 1900-1901. 4 vols.

b. *Atti del Congresso Internazionale dei Fisici; Como, Pavia, Roma, 1927.* Bologna: Nicola Zanichelli, 1928. 2 vols.

c. *International Conference on Physics, London, 1934. Papers and Discussions.* Cambridge, England: At the University Press, 1935. 2 vols. (Vol. 1: Nuclear physics; Vol. 2: Solid state of matter.)

d. *International Conference of Theoretical Physics; Kyoto and Tokyo, 1953. Proceedings.* Tokyo: Science Council of Japan, 1955. 942 pp.

e. *International Nuclear Physics Conference; Gatlinburg, Tenn., 1966. Proceedings.* New York: Academic Press, 1967. 1118 pp.

Other conferences that have issued proceedings include the Congresses for Applied Mechanics; Conferences on Spectroscopy; National Electronics Conferences; Conferences on High-Energy Physics; etc.

Those interested in planning and conducting a conference may refer to the following for guidance:

Kindler, Herbert S. *Organizing the Technical Conference.* New York: Reinhold Publishing Corporation, 1960. 139 pp.

Government publications.

The United States Government issues several scientific periodicals, as well as an abundance of separate materials appearing irregularly in pseudo-serial form. The following will be found most helpful:

a. Boyd, Anne M., and Rips, Rae E. *United States Government Publications.* (Third Edition.) New York: The H. W. Wilson Company, 1949. 627 pp.

b. Schmeckebier, Laurance F., and Eastin, R. B. *Government Publications and Their Use.* (Revised Edition.) Washington, D. C.: The Brookings Institution, 1961. 476 pp.

National Bureau of Standards

Of physics interest are the U. S. National Bureau of Standards' publications, notably the monthly *Journal of Research* now comprised of three separate parts: A, Physics and chemistry; B, Mathematics and mathematical physics; and C, Engineering and instrumentation. The *Technical News Bulletin* describes work in progress. There are many *Circulars* (discontinued in 1959), *Miscellaneous Publications, Monographs,* etc., on a wide range of topics. The Bureau issues checklists [11] of its prior publications at intervals, and annual reports on its diversified activities. It is no longer located in Washington, D. C.[12] For an account of its development see:

Measures for Progress: A History of the National Bureau of Standards, by Rexmond C. Cochran. 1966. 703 pp. (NBS Misc. Publ. No. 275.)

For further discussion of standards in general, see Chapter V.

Atomic Energy Commission

Another governmental agency vitally linked with physics research and experimentation is the United States Atomic Energy Commission.[13] Besides reports, it issues the *Technical Progress Reviews,* a series of four quarterlies on nuclear technology: 1, Isotopes and radia-

---

11. See its *Circular* No. 460 and supplement, for 1901 to mid-1957; and *Misc. Publ.* No. 240 and supplement, for mid-1957 to mid-1966.
12. See G. B. Lubkin, "NBS Moves to Gaithersburg [Md.]" *Physics Today,* 19 (No. 11): 36-42, November 1966.
13. See D. E. Gray, "Dissemination of Technical Information by AEC." *Physics Today,* 4 (No. 11): 22-24, November 1951.

tion technology; 2, Nuclear safety; 3, Power reactor technology; and 4, Reactor materials.

During the period following the second world war, the *National Nuclear Energy Series* was published by the McGraw-Hill Book Company as a record of the research work done under the "Manhattan Project" and later under the Atomic Energy Commission.

Nuclear Science Abstracts: Non-classified material appears in *Nuclear Science Abstracts,* issued semi-monthly by the Atomic Energy Commission, and "abstracting as completely and as promptly as possible the literature of nuclear science and engineering." (Its predecessor was *Abstracts of Declassified Documents,* vols. 1-2, July 1947-June 1948, with combined index.) Author, subject and report-number indexes to individual issues are cumulated at intervals during the year, and resultant annual indexes are combined quinquennially. Abstracts give journal citations, and the numbers of reports. Report numbers may be identified in the latest cumulated numerical index and its supplements, which disclose source and price. *New Nuclear Data Cumulations* were included 1952-1957; for continuations see index of this *Guide* under Nuclear data.

Public availability of USAEC reports: 1, At USAEC depository libraries; 2, In scientific and technical journals, etc.; 3, From Microcard Editions, Inc. (West Salem, Wisc. 54669); 4, From Clearinghouse for Federal Scientific and Technical Information; and 5, From the Superintendent of Documents.

Reports

The current title of the government abstracting service for scientific and technical reports [14] is *U. S. Government Research and Development Reports,* with its companion comprehensive . . . *Index,* both being published semi-monthly by an NBS agency, the Clearinghouse for Federal Scientific and Technical Information. During the period 1946 through 1964, the Office of Technical Services had performed the function under three successive titles: 1, *Bibliography of Scientific and Industrial Reports;* 2, *Bibliography of Technical Reports;* and 3, *U. S. Government Research Reports.* Timesaving multi-year numerical and correlation indexes stem from the Special Libraries Association (1946-1948), and from Technical Information Service (1949-1954; 1955-1964; and 1946-1967). Correlation of report numbers is further facilitated by another S.L.A.

14. See R. E. Burton and B. A. Green, Jr., "Technical Reports in Physics Literature." *Physics Today,* 14 (No. 10): 35-37, October 1961.

Correlation Index (1946-1952). These and various yearly indexes are fully described in:

Boylan, Nancy G. "Identifying Technical Reports Through U. S. Government Research Reports and Its Published Indexes." *College and Research Libraries,* 28: 175-183, May 1967.
A chart (page 181) shows which indexes to use for each volume of the 1946-1964 series.

For help with code numbering see a *Dictionary of Report Series Codes* (S.L.A., 1962).

Twelve Federal Regional Technical Report Centers are in operation throughout the country to facilitate access to unclassified AEC, NASA, and other reports. One is at Columbia University's Engineering Library; another at John Crerar Library, Chicago.

See also index under Reports, technical.

Patents

Patent literature is also of physics interest.[15] The United States Patent Office distributes patent descriptions called specifications, that are abstracted in the weekly *Official Gazette,* and are finally listed in the *Annual Index,* formerly by catch-title, but now only under Patentees, with an appended patent-classification finding scheme. Copies of patents may be purchased at fifty cents from the Commissioner of Patents, Washington, D. C., or examined at certain libraries.

Several volumes of an extensive patent-indexing program undertaken by Rowman and Littlefield, Inc., have appeared, but the project is temporarily suspended. It comprises the *National Catalog of Patents* and the *Index of Patents 1790-1960.* The former features claims and drawings from *Official Gazette* entries, in classified array. The latter's pages display pertinent patent numbers according to the classification scheme, with cross-references. Chemical and electrical sections are now available.

Lists

These bibliographical tools are useful for finding references to government documents:

a. *Document Catalog.* This comprehensive analytical catalog by

---

15. See also the following: G. M. Naimark, *A Patent Manual for Scientists and Engineers.* Springfield, Ill.: Charles C. Thomas, 1961. 108 pp.; two U. S. Patent Office brochures entitled *General Information Concerning Patents* and *Patents and Inventions;* Radzinsky's *Making Patent Drawings;* and Houghton's *Technical Information Sources.*

subject, author and department was issued biennially from the 53rd through 76th Congress, and covers 1893-1940.

b. *Monthly Catalog*. One must consult it for the period since 1940 as a useful although briefer listing of the majority of government documents. Decennial indexes eventually have appeared: 1941-1950 in 1953; 1951-1960 in 1967.

c. *List of Selected United States Government Publications for Sale by the Superintendent of Documents*. This lists a few new items of wide appeal, bi-weekly.

d. *Price Lists*. These itemize only documents still in print and on sale, for such areas as Atomic energy in No. 84; Chemistry in No. 46; Electricity and electronics in No. 82; Physics, and scientific standards or tests in No. 64; and Space programs in No. 79A.

e. *Checklists*. Agencies often issue indexes of their own publications.

Symbols indicating availability of government publications are explained at the bottom of *Monthly Catalog* pages, *e.g.*, the Greek letter Phi means for sale by the Clearinghouse for Federal Scientific and Technical Information; asterisks denote items sold by the Superintendent of Documents; single daggers direct inquiry to the issuing agencies; and large black dots indicate availability at depository libraries.

Using periodicals.

Although random scanning of journals sometimes yields relevant material, consulting the abstracts and indexes previously mentioned is a far more efficient procedure. Familiarity with subject heading techniques is of prime importance, and may be developed by: (1) acquisition of sufficient subject background for orientation purposes; and (2) examination and comparison of headings used in periodical indexing media. These indexes choose particular terms under which to list articles, with cross references from synonymous and related terms. Unfortunately the choices vary among different media, as tabulated elsewhere.[16]

After doing some preliminary reading on a subject, one should prepare a list of likely headings to be searched, adding others as required. It is preferable to work from current volumes of indexes

16. R. H. Whitford and J. B. O'Farrell, "Use of a Technical Library." *Mechanical Engineering*, 70: 987-993, December 1948, especially pp. 989-990.

towards earlier ones, for this practice places emphasis upon recent material and may be conveniently stopped whenever enough items have been garnered, or when a comprehensive review article adquately summarizing prior research has been discovered. Attention should be given to all "*see*" references (from inactive to chosen headings), and "*see also*" references (among related headings). Proceeding systematically and accurately, record selected entries on separate slips in the following manner for most general [17] purposes:

Weinstein, M. A.—Magnetic decay in a hollow elliptic cylinder. J. Appl. Phys., 37: 248-253, Jan. 1966. (N.B.: "37: 248-253" means volume 37, pages 248-253 inclusive.) [18]

A *book* entry might be:

Freeman, Ira M.—Modern introductory physics, 2nd ed., N. Y., McGraw-Hill, 1957. 497 p.

2—Books

The separate publications known as books differ widely in size, shape, format, purpose, level of understanding, and comprehensiveness.

Some serve their current purposes but soon are superseded and forgotten. Others survive because of special merit and lasting utility, becoming known as classics:

> What is a classic? It is an enduring work of excellence and authority. It can be a painting, a symphony or a novel. *It can be a work of science or engineering, too.*
>
> Potential classics in science and engineering are being written today. Time alone can tell which of them will endure. Surely, they will be found among the books which are *today* accepted as leading authorities in their fields.[19]

Certain worthwhile out-of-print books for which there is continuing demand are reissued photographically by Dover Publications, Inc.; University Microfilms, A Xerox Company; Chelsea Publishing Company; Johnson Reprint Corporation; Harper and Row; Hafner Publishing Company; etc.

See also index under Source material.

---

17. For publishers' style requirements, however, see index under Technical Writing.
18. A few journals page each issue separately, in which case it is desirable to indicate issue number also; e.g., *Physics Today*, 1 (No. 8): 6-14, December 1948.
19. From Bell Telephone Laboratories insert. *Physics Today*, 12 (No. 12): 3, December 1959.

General.

As previously noted, books do not contain the latest work in any field, for they are selective summaries that may be two or three years behind current practice, even at publication time. They range from multi-volume encyclopedias to introductory texts, and cover many different aspects (historical, mathematical, educational, topical, etc.). Very frequently they are published as units of a *series,*[20] for example:

a. *Methuen's Monographs on Physical Subjects.* London: Methuen and Company.
These are pocket-size compendia noted for clarity of presentation.

b. *International Series in Pure and Applied Physics.* New York: McGraw-Hill Book Company.

c. *Cambridge Monographs on Physics.* Cambridge, England: At the University Press.

d. *Interscience Monographs and Texts in Physics and Astronomy.* New York: Interscience Publishers, Division of John Wiley and Sons.

e. *Pure and Applied Physics: A Series of Monographs and Textbooks.* New York: Academic Press.

f. *International Series of Monographs on Physics.* Oxford: At the Clarendon Press.
Advanced treatises for specialists.

g. *International Series of Monographs in Nuclear Energy.* New York, Oxford, etc.: Pergamon Press.

h. *Momentum Books,* published for the Commission on College Physics. Princeton, N. J.: D. Van Nostrand Company.

i. *Science Study Series.* Garden City, N. Y.: Doubleday and Company; London: William Heinemann. These stem from the Physical Science Study Committee (PSSC), and bridge the gap between scientist and layman.

j. *Life Science Library.* New York: Time, Inc. Colorful books in popular style by authorities.

The foregoing series have been selected from among many possible examples on the basis of their diversified character. Advanced textbooks and treatises represented therein cannot be sharply differentiated, except that the former tend to be introductory teaching presentations,

---

20. Most publishers' series of discrete books are usually scattered by subject throughout a library collection. A series note on catalog cards serves to locate units of series that are shelved together, e.g., the *M.I.T. Radiation Laboratory Series,* an integral set with overall index volume.

while the latter are well-documented comprehensive records possessing reference utility.

John Wiley's E. P. Hamilton indicates a distinction between successive "editions" and "printings" that should be more universally observed:

> In the earlier days of publishing, the term "new edition" seems to have been used carelessly. Often a so-called new edition amounted to little more than a new printing, perhaps with minor corrections. Today we insist that a new edition means a really substantial revision. Some authors think that if they add a chapter and make a few corrections elsewhere their book might qualify as a new edition. With this viewpoint we disagree. We prefer to call such a book a corrected printing, and are willing to add a line on the title page to the effect that a new chapter has been included.[21]

The science editor at McGraw-Hill takes a look at book costs:

> Broad, diverse and worrisome problems are affecting the economics of scientific and technical book publishing: such things as the cost of book manufacturing, effects of technical innovations such as photocopying machinery and computerized books, increased involvement of government agencies in publishing and curriculum revision, and the concept of "fair use" of copyrighted material. The publisher finds himself in a dilemma, caught between the consumer who asks why some books cost so much and a decreasing market for those same books that pushes their cost even higher.[22]

However, as optimistically reported in *Physics Today,* 20 (No. 10): 79-81, October 1967, "greater student enrollments, expanded U. S. and overseas markets and the general inflation make the physics book market a healthy and growing one" despite costs and competition.

Bibliographies.

The term "bibliography" may be taken to denote any list of books and other printed materials, thus including lists associated with the publishing trade as well as those usually appended to research studies. A subject bibliography may be selective or comprehensive within set limits, and is most valuable when annotated and arranged by topical subdivision, although chronological and author arrangements are also employed.

---

21. E. P. Hamilton, "Engineering Literature." *Library Journal,* 77: 1939-1941, November 15, 1952.
22. T. J. Dembofsky, "Why Technical Books Cost What They Do." *Physics Today,* 19 (No. 3): 65-67, March 1966.

An example of a bibliography of bibliographies [23] is:

Darrow, Karl K. *Classified List of Published Bibliographies in Physics 1910-1922.* Washington, D. C.: National Research Council, 1924. 102 pp. (Its *Bulletin,* Vol. 8, Pt. 5, No. 47, July 1924.)

Guides to reference books enable one to find physics material in the more general bibliographical tools. The following will prove extremely helpful on many occasions:

a. Winchell, Constance M. *Guide to Reference Books.* (Eighth Edition.) Chicago: American Library Association, 1967. 741 pp. The outstanding guide to reference books basic to research. Arranged in five parts: 1, General reference works; 2, Humanities; 3, Social sciences; 4, History and area studies; and 5, Pure and applied sciences. Grouping under subject divisions is by type of reference book, e.g., guides, bibliographies, indexes, encyclopedias, etc. Includes books published through 1964; supplemented informally by "Selected Reference Books" appearing in January and July issues of *College and Research Libraries* as a project of the Reference Department of the Columbia University Libraries.

b. Roberts, Arthur D. *Introduction to Reference Books.* (Third Edition.) London: The Library Association, 1956. 237 pp.

Special subject guides provide overall surveys, search techniques, and specific sources. Another useful guide for physics *per se* has made its appearance:

Yates, B. *How to Find Out about Physics; A Guide to Sources of Information, Arranged by the Decimal Classification.* Oxford, London, etc.: Pergamon Press, 1965. 175 pp.
This is "selective and includes material on careers, societies and non-literature sources of information," as well as chapters on various forms of literature, the subdivisions of physics, etc.

Guides to both physics and related fields include:

a. Parke, Nathan G. *Guide to the Literature of Mathematics and Physics, Including Related Works on Engineering Science.* (Second Edition.) New York: Dover Publications, 1958. 436 pp.
There are over 5000 entries, with "selection biased in the direction of applied mathematics." Chapters on searching book and periodical literature are followed by an extensive bibliography arranged by the

---

23. See also T. Besterman, *A World Bibliography of Bibliographies.* (Fourth Edition.) Lausanne, Switzerland: Societas Bibliographica, 1965-1966. 5 vols. *Bibliographic Index* was cited earlier.

topics of physics, mathematics and engineering within a single alphabet.

b. Walford, Albert J., editor. *Guide to Reference Material.* (Second Edition.) Vol. 1: *Science and Technology.* London: The Library Association, 1966. 483 pp.

In attractive format, this work "provides a signpost to reference books and bibliographies published mainly in recent years, international in scope, but with emphasis on items published in Britain." In the present volume see especially the chapters "Science and technology" and "Physics." (Volumes 2 and 3 will deal with social and historical sciences, literature, fine arts, etc.)

c. Jenkins, Frances B. *Science Reference Sources.* (Fourth Edition.) Champaign, Illinois: Illini Union Bookstore, 1965. 143 pp.

For the various sciences, reference materials are grouped by form, e.g., guides, bibliographies, indexes, reviews, histories, handbooks, serials, etc.

d. Johnson, Irma. *Selected Books and Journals in Science and Engineering.* (Second Edition.) Cambridge, Mass.: The M.I.T. Press, 1959. 63 pp.

This is a guide to first purchases in science and engineering for undergraduate libraries.

e. Malclès, Louise-Noëlle. *Les Sources du Travail Bibliographique.* Vol. 3: *Bibliographies Spécialisées (Sciences Exactes at Techniques).* Geneva: Droz, 1958. 575 pp.

Constitutes a most comprehensive subject-guide.

f. Malinowsky, Harold R. *Science and Engineering Reference Sources; A Guide for Students and Librarians.* Rochester, N. Y.: Libraries Unlimited, 1967. 213 pp.

The foregoing may well be used in conjunction with the present *Guide,* supplementing one another. Other reference manuals are listed in the General Bibliography, appended.

A selective listing of American books published 1930-1956 is:

Hawkins, Reginald R., editor. *Scientific, Medical, and Technical Books Published in the United States of America: A Selected List of Titles in Print.* (Second Edition: Books Published to December 1956.) Washington, D. C.: National Research Council, 1958. 1491 pp.

The first edition and its two supplements were for 1930-1952.

The British counterpart is issued by the Association of Special Libraries and Information Bureaux:

British Scientific and Technical Books, 1935-1952. London: ASLIB, 1956. 364 pp. . . . *1953-1957.* ASLIB, 1960. 251 pp. . . . 1958-
Its predecessor was the British Science Guild's non-selective *Catalogue* . . . (3rd ed., 1930. 754 pp.) For standard works in print ASLIB publishes at intervals a *Select List* . . . (5th ed., 1957. 88 pp.)

Books in the English language [24] (no matter where published) on any subject may be found in the *Cumulative Book Index,*[25] issued monthly by the H. W. Wilson Company (in continuation of its *United States Catalog* series). Whether a book originally listed is still obtainable may be ascertained from *Books in Print* [26] which is an author-title-series index to *Publishers' Trade List Annual,* a collection of the catalogs of American publishers. There is also a useful companion index, *Subject Guide to Books in Print.*

Bi-monthly *Forthcoming Books* not only forecasts books, but cumulates those published since the last *Books in Print* appeared. It is paralleled by *Subject Guide to Forthcoming Books.*

Four volumes of *American Scientific Books* published during the period 1960-1965 had been compiled by Phyllis B. Steckler from the monthly *American Book Publishing Record.*[27] Henceforth annual cumulations from the latter will list the scientific books in decimal location among all publications, rather than separately.

Current annotated lists and reviews enable one to appraise publications. Entries in *Publishers' Weekly,* though brief, identify new books of scientific publishers in this country. *New Technical Books* is a monthly periodical describing recent acquisitions of the New York Public Library's Science-Technology Department. Books of wide appeal may receive entry in *Booklist and Subscription Books Bulletin,* which has included evaluation of publishers' sets since the 1956 merger. Read also entries in *Choice* (published monthly by the Association of College and Research Libraries), and in the review columns of *Library Journal.* Selected British publications appear in the monthlies, *British Book News* and *ASLIB Booklist;* all are listed in the weekly *British National Bibliography.* To keep thoroughly informed, one should scan the book reviews that appear in technical

---

24. But use also the cumulative *British National Bibliography.*
25. British counterpart: *Whitaker's Cumulative Book List.*
26. Comparable with *British Books in Print: The Reference Catalogue of Current Literature,* published by J. Whitaker and Sons, Ltd. This firm also began *Technical Books in Print* in 1964.
27. Cumulated from the "Weekly Record" listings of *Publishers' Weekly.*

journals, notably *American Journal of Physics, Review of Scientific Instruments, Physics Today, Science News, Nature, Science,* etc. (After an appreciable time lapse, excerpts from reviews appear in *Technical Book Review Index,* which also serves as a new-book check-list.)

Many sources of foreign publications, such as the *Deutsche Bibliographie,* and the French *Biblio* and *Bibliographie de la France,* are listed in Winchell. (For translated works, see UNESCO's *Index Translationum.*)

Examples of special-purpose bibliographies are:

a. *McGraw-Hill Basic Bibliography of Science and Technology: Recent Titles on More Than 7000 Subjects,* compiled and annotated by the editors of the *McGraw-Hill Encyclopedia of Science and Technology.* New York: McGraw-Hill Book Company, 1966. 738 pp. Bibliographical extension of the articles, under corresponding headings.

b. Northeastern University Library Staff, compilers. *A Selective Bibliography in Science and Engineering.* Boston: G. K. Hall and Company, 1964. 550 pp.
"Not intended as a list of the best books . . . but rather a working collection . . . for an undergraduate library."

c. Voigt, Melvin J., and Treyz, Joseph H. *Books for College Libraries;* A selected list of approximately 53,400 titles based on the initial selection made for the University of California's New Campuses Program, and selected with the assistance of college teachers, librarians, and other advisers. Chicago: American Library Association, 1967. 1056 pp. "Not . . . a list of the best books or a basic list for any college library, for selection of books for a college library must be made in terms of the needs of that particular institution." This retrospective book selection aid includes only titles published prior to 1964, and is intended to update the following list:

Shaw, Charles B., compiler. *A List of Books for College Libraries.* (Second Preliminary Edition.) Chicago: American Library Association, 1931. 810 pp.; *Supplement, 1931-1938.* Chicago: The Association, 1940. 284 pp.

The *Resource Letters* published in the *American Journal of Physics* and separately are described in Chapter VII. (*See index* under Source material, teaching.) The following one is of bibliographical interest:

Bork, Alfred M., and Arons, Arnold B. "Collateral Reading for Physics Courses." *American Journal of Physics,* 35: 71-78, February 1967. (Resource Letter ColR-1.)

For bibliography of early works one may turn to:

Young, Thomas. *A Course of Lectures on Natural Philosophy and the Mechanical Arts.* London: Joseph Johnson, 1807. 2 vols.
This first edition includes (Vol. 2, pp. 87-520) a selective subject bibliography of 20,000 book and periodical references systematically collected through 1805. It was omitted in a later edition dated 1845, but is still valuable. Crew declares:

> No serious minded student of physics will fail to make the acquaintance of Dr. Thomas Young; for he is the last man, unless one includes Whewell and Helmholtz, of the race that knew everything that was to be known—the group that includes Aristotle, Leonardo da Vinci, Stevin, Galileo, Descartes, Huygens, Franklin.[28]

A fine limited edition constitutes a listing of outstanding classics, with facsimile pages therefrom:

Horblit, Harrison D. *One Hundred Books Famous in Science,* based on an exhibition held at the Grolier Club in 1958. New York: The Grolier Club, 1964. 449 pp.

Old works, as they pass through book auctions, may be listed in *American Book-Prices Current* which has collective indexes. The Henry Sotheran Company's *Bibliotheca Chemico-Mathematica* is a pretentious multi-volume catalog of scientific classics, with interesting comments on their contents and special features. Extensive card catalogs of long-established libraries may be consulted for older printed material, especially when books are chronologically grouped for major subjects. Finally, one may use earlier bibliographies, such as:

John Crerar Library. *A List of Bibliographies of Special Subjects.* Chicago: The Library, 1902. 504 pp.

Dissertations.

There are guides to lists of theses:

a. Palfrey, Thomas R., and Coleman, Henry E. *Guide to Bibliographies of Theses, United States and Canada.*[29] (Second Edition.) Chicago: American Library Association, 1940. 54 pp.
Its three parts are: General lists of dissertations in all fields; Lists in special fields; Institutional lists.

---

28. H. Crew, *The Rise of Modern Physics,* pp. 245-246. (Second Edition.) Baltimore: Williams and Wilkins Company, 1935. See also: Alexander Wood, *Thomas Young* . . . Cambridge, England: At the University Press, 1954. 355 pp.
29. For additions and corrections, see R. P. Rosenberg, "Bibliographies of Theses in America." *Bulletin of Bibliography,* 18: 181-182 and 201-203, September-December 1945 and January-April 1946.

b. Black, Dorothy M. *Guide to Lists of Master's Theses.* Chicago: American Library Association, 1966. 144 pp.

One finds physics dissertations in the following, depending on date:

a. Marckworth, M. Lois, *et al.*, compilers. *Dissertations in Physics: An Indexed Bibliography of All Doctoral Theses Accepted by American Universities, 1861-1959.* Stanford, Cal.: Stanford University Press, 1961. 803 pp.

b. *Doctoral Dissertations Accepted by American Universities, 1933/34-1954/55.* Compiled for the Association of Research Libraries. New York: The H. W. Wilson Company, 1934-1956. 22 vols.

c. *Index to American Doctoral Dissertations,* 1955/56—Compiled for the Association of Research Libraries. Ann Arbor, Mich.: University Microfilms, 1957-
Issued as the final and index issue of each volume of *Dissertation Abstracts,* described below.

For science theses not found in Marckworth, use the preceding lists or the following one:

U. S. Library of Congress. *List of American Doctoral Dissertations Printed in 1912 to 1938.* Washington, D. C.: Government Printing Office, 1913-1940. 26 vols.
In this discontinued record, emphasis was placed upon printed form rather than mere acceptance.

In addition, one should consult the Palfrey and Coleman section on "Sciences" for the rather complicated array of acceptances for the period 1898-1933 during which listing occurred in *Science, School and Society,* and the National Research Council's *Reprint and Circular Series.*

Dissertation Abstracts (formerly *Microfilm Abstracts*) lists dissertations available in microfilm or paper form through University Microfilms, A Xerox Company. Beginning July 1966 it is being published in two monthly sections: (A) covering the humanities and social sciences; and (B) for the physical sciences and engineering. See its combined *Index,* above. (Paired with this is a computerized *Datrix* service which retrieves dissertation titles pertinent to keyword inquiries.)

Masters theses are listed in:

a. Bledsoe, Barton. *Masters Theses in Science, 1952.* Washington, D. C.: Biblio Press, 1954. 252 pp.
The only volume of a proposed annual list.

b. *Masters Theses in the Pure and Applied Sciences Accepted by*

Colleges and Universities of the United States. Vol. 1—1955/56— Lafayette, Ind.: Purdue University, Thermophysical Properties Research Center, 1957-

c. *Masters Abstracts: Abstracts of Selected Masters Theses on Microfilm.* Ann Arbor, Mich.: University Microfilms, 1962- Quarterly.

Using books and libraries.

Books are arranged on library shelves according to a classification scheme (Dewey, Library of Congress, etc.) so as to group books on a subject together, as well as in relative sequence. Call numbers are printed on books and catalog cards for identification and shelving purposes. The library's card catalog serves as an index to its book holdings, but must be supplemented with published indexes that analyze periodical contents, etc. Its form is usually "dictionary," i.e., subject,[30] author and title cards in one alphabet, but some libraries favor separated or classified files. It should be noted that books are normally given an added entry under title only when it is distinctive, e.g., *Hidden Hunger* (a book on vitamins). Cards bear standardized bibliographical information, and are alphabetized according to definite filing rules with respect to word by word sequence (electron optics before electronics); omission of initial articles; grouping of *Mac* . . . variants, etc. Helpful clues to further literature are provided in the form of bibliography notes and cross reference cards. General reference works should not be overlooked.

To reveal more material beyond local holdings, such tools as the *Cumulative Book Index* may be consulted, or possibly the published catalogs of other libraries (Engineering Societies, John Crerar, University of California, etc.), as well as Library of Congress [31] and British Museum catalogs continuing in printed form. Under certain restrictions, books may sometimes be examined at or borrowed [32] from nearby libraries.

Resources are indicated in:

a. Ash, Lee. *Subject Collections: A Guide to Special Book Collec-*

---

30. Subject headings are standardized, possibly according to the Library of Congress List, supplemented by M. J. Voigt, *Subject Headings in Physics.* Chicago: American Library Association, 1944. 151 pp.

31. Beginning 1956: *National Union Catalog,* including LC printed catalog cards and other entries.

32. See "General Interlibrary Loan Code 1952." *College and Research Libraries,* 13: 350-358, October 1952.

tions and Subject Emphases as Reported by University, College, Public and Special Libraries in the United States and Canada. (Third Edition.) New York: R. R. Bowker Company, 1967. 1221 pp.

b. *Special Library Resources.*[33] New York: Special Libraries Association, 1941-1947. 4 vols.

c. National Referral Center for Science and Technology. *Directory of Information Resources in the United States: Physical Sciences, Biological Sciences, Engineering.* Washington, D. C.: Government Printing Office, 1965. 352 pp.

d. Organization for European Co-Operation and Development. *Guide to European Sources of Technical Information.* (Second Edition.) Paris: OECD, 1964. 292 pp.

See also "Toward National Information Networks," in *Physics Today* 19 (No. 1): 38-60, January 1966.

See also index under Library resources.

Periodical article reprints are often available from the author or his affiliated group, and photocopies of books and articles are obtainable from sources listed in:

Raymond, J. G. *Directory of Microfilm Services in the United States and Canada.* (Revised Edition.) New York: Special Libraries Association, 1947. 30 pp.

See also indications of services of coöperating libraries, preceding the 1961 *Chemical Abstracts* periodicals list; and the University Microfilms *OP Catalog.*

Some libraries will even, for nominal fees, prepare translations, bibliographies, abstracts, etc. Further material on efficient use of library collections and facilities may be found among the books listed in the General Bibliography.

Dwight E. Gray summarizes the development of science-related activities in our national library under the title "Science, Technology, and the Library of Congress," in *Physics Today,* 18 (No. 6): 44-48, June 1965.

Libraries are undergoing change with the adoption of new information retrieval ideas, outlined in:

a. Vickery, B. C. *On Retrieval System Theory.* (Second Edition.) London: Butterworths Scientific Publications, 1965. 191 pp.

b. Kent, Allen. *Textbook on Mechanized Information Retrieval.* (Second Edition.) New York: Interscience Publishers, 1966. 371 pp.

---

33. See also the latest *Special Libraries Directory* published by the Association.

c. Perry, James W.; Kent, Allen; and Berry, Madeline M. *Machine Literature Searching*. New York: Interscience Publishers, 1956. 174 pp.

Before bibliographical assistance to scientists can be improved, however, their present information techniques should be observed:

Voigt, Melvin J. *Scientists' Approaches to Information.* Chicago: American Library Association, 1961. 81 pp. (ACRL Monograph No. 24.)

Three basic motives are noted (p. 21):

> This need to keep up to date with the current progress of a scientist's field is called the *current approach.* . . . The need for specific information directly connected with the research work or the problems at hand is called the *everyday approach.* . . . The requirement for all information, or as much as can be found, can be called the *exhaustive approach.*

For physics literature approaches see pp. 65-73 of this monograph.

CHAPTER III

HISTORICAL APPROACH

History and Current Events

Whenever information concerning past events is sought, the "Historical Approach" is indicated. One consults retrospective accounts of the development of physics among the natural sciences, or turns to more specialized versions. Current summaries are also available.

Intrinsically, any historical treatment embraces considerable biographical data. Accordingly, the books cited may be used as sources of such information by scanning their indexes. For topics of inquiry distinctly biographical in flavor rather than chronological, however, see Chapter IV.

Overview

Physics had its origins in the early Greek schools of thought which were devoted to the description of the fundamental nature and substance of the universe. These explanations, although fanciful and imaginative, have been credited with harboring the germ of the modern atomic and other theories. However, the beginning of physics as an experimental science is conceded to date from the close of the sixteenth century, when Galileo Galilei dared to base his conclusions on observed data instead of Aristotelian intuitive dicta. The succeeding centuries yielded great cumulative advance in physics under the all-embracing appellation, "Natural Philosophy," for only comparatively recently has physics been so known. The name modern physics (as distinguished from classical physics) is usually applied to the new atomic physics which began with the discovery of X-rays by Roentgen in 1895 and received great impetus from Planck's proposal of discrete quanta of radiation four years afterwards.

Without an adequate understanding of the past, the present cannot be fully appreciated. Intelligent grasp of today's knowledge in any field depends largely on a realization of the multifarious paths along which progress was made, the interrelationships of diverse lines of activity, and the gradual synthesis of successive findings and achievements. As Chalmers states:

> The historical approach towards any branch of scientific study possesses not only intrinsic interest; as a practical educative method it undoubtedly

has distinct merits. It provides the easiest path towards the ready understanding of modern knowledge and opinion. However involved and abstruse any subject may appear in its modern dressing, it remains true in every case that it has been built up from simple foundations, from elementary observations made in past generations and the natural cogitations which they inspired in successive minds.[1]

The late Lloyd W. Taylor [2] was a foremost proponent of the idea that textbooks designed for regular courses in college physics should feature considerable historical material. Unfortunately, the only students whose thoughts are directed along historical lines seem to be the non-physics majors, via their "science survey" courses. Blüh maintains that *all* need such enlightenment:

A modern Erasmus, in a present-day *In Praise of Folly,* would probably feel inclined to direct some of his critical remarks against men of science who demonstrate forcefully the necessity of a scientific education for everyone, but seem to be blind to the equal need of a broad education for themselves. The historical approach, for example, so widely advocated and propagated in the teaching of natural science as a cultural element in general education is only applied to the nonscientist, while the professional education of the scientist makes no effort to concern itself with historical considerations.[3]

I. B. Cohen urges use of interesting historical material and cites many sources in his article, "A Sense of History in Science" (*American Journal of Physics,* 18: 343-359, September 1950).

Teaching objectives that link history, philosophy and science are delineated in:

Seeger, Raymond J. "On Teaching the History of Physics." *American Journal of Physics,* 32: 619-625, August 1964.

Guides

The reading of historical accounts becomes more meaningful when one is acquainted with compilation methodology, as related by a master chronicler:

Sarton, George. *A Guide to the History of Science.* Waltham, Mass.: Chronica Botanica Company, 1952. 316 pp.

Three introductory essays on "Science and Tradition" are followed by an extremely useful literature guide, expanded from that earlier

1. T. W. Chalmers, *Historic Researches,* p. 1. New York: Charles Scribner's Sons, 1952.
2. See his *Physics; the Pioneer Science.*
3. O. Blüh, "The History of Physics and the Old Humanism." *American Journal of Physics,* 18: 308-311, May 1950.

appended to the author's *Study of the History of Science* (1936). Some of the sections of this 236-page bibliography are: Scientific methods and philosophy of science; Science and society; National academies and national scientific societies; Treatises and handbooks on the history of science; History of special sciences; Journals and serials concerning the history and philosophy of science; Institutes, museums, libraries; International congresses.

Chronologies.

A most comprehensive chronological arrangement of 13,000 scientific discoveries and inventions from 3500 B.C. to the year of publication is:

Darmstaedter, Ludwig. *Handbuch zur Geschichte der Naturwissenschaften und der Technik in Chronologischer Darstellung.* (Zweite Auflage.) Berlin: Springer, 1908. 1262 pp. There are name and subject indexes to the brief descriptive items.

Another useful set of chronological tables is:

Auerbach, Felix. *Geschichtstafeln der Physik.* Leipzig: J. A. Barth, 1910. 150 pp.

Its four parts are: (1) Chronological table from 650 B.C. to 1900 A.D. giving important scientific discoveries, with year and physicist's name; (2) Chronological table of important physics books with year, author and place of publication from 350 B.C. to 1900 A.D.; (3) Chronological table of physicists, with dates of birth and death, from Pythagoras (560 to 490 B.C.) to Drude (1863-1906); and (4) Alphabetical index to the first part. Note that a later book[4] by Auerbach carries his tables of scientific discoveries and physicists through 1923.

A contemporary chronology:

U. S. National Aeronautics and Space Administration. *Aeronautics and Astronautics; An American Chronology of Science and Technology in the Exploration of Space, 1915-1960,* by Eugene M. Emme. Washington, D. C.: NASA, 1961. 240 pp.

Arranged by dates of events. Supplements issued: *Astronautical and Aeronautical Events of* 1961-

Chronological tables are often appended to historical treatments, as noted throughout this chapter.

---

4. F. Auerbach, *Entwicklungsgeschichte der Modernen Physik.* Berlin: Springer, 1923. 344 pp.

Bibliographies.

Relevant books in important library collections are listed in:

a. John Crerar Library. *A List of Books on the History of Science.* Chicago: The Library, 1911. 297 pp.

First Supplement: 1917. 139 pp.

Second Supplement: Parts I-VI (1942-1946), of which Part IV (1944) is for Physics, 12 pp.

b. Oklahoma University Library. *Check List of the E. De Golyer Collection in the History of Science and Technology,* compiled by Arthur McAnally and Duane H. D. Roller. (Third Edition.) Norman, Okla.: University of Oklahoma Press, 1954. 127 pp.

Other extensive bibliographical lists on the history of science may be found in Sarton's *A Guide to the History of Science,* noted before; and appended to the books by Pledge and Sedgwick cited under Science Histories, below.

In a particular library, historical treatments may be found under the headings "Physics—History" and "Science—History" in the card catalog. New books are listed in the *Cumulative Book Index.*

Historical Accounts

Books dealing exclusively with physics are of greatest interest to the users of this guide. However, it will be found necessary to supplement the relatively short histories of physics with the longer works available on the general history of science.

Physics histories.

One may start with readable surveys:

a. March, Arthur, and Freeman, Ira M. *New World of Physics.* New York: Random House, 1962. 195 pp.
Development of physical theory since Aristotle's era.

b. Schneer, Cecil J. *Search for Order: The Development of the Major Ideas in the Physical Sciences from the Earliest Times to the Present.* New York: Harper and Row, 1960. 398 pp.

c. Amaldi, Ginestra. *The Nature of Matter; Physical Theory from Thales to Fermi.* Chicago: University of Chicago Press, 1966. 332 pp.

d. Toulmin, Stephan, and Goodfield, June. *The Architecture of Matter.* New York: Harper and Row, 1962. 399 pp.
An interesting unified history of thoughts on matter from Ionian times to modern.

e. Chalmers, Thomas W. *Historic Researches; Chapters in the History of Physical and Chemical Discovery.* New York: Charles Scribner's Sons, 1952. 288 pp.
The chapters report research concerning friction, mechanical equivalent of heat, electro-dynamics, molecular physics, conduction of electricity through liquids and gases, X-rays, etc. They show "not only who made the discovery or formulated the theory—but how he came to make it and against what background of contemporary thought and achievement it was made." Biographical sketches appear on pp. 259-274. *Cf.* M. H. Shamos' *Great Experiments in Physics.*

Less recent but well-known presentations are:

a. Crew, Henry. *The Rise of Modern Physics.* (Second Edition.) Baltimore: Williams and Wilkins Company, 1935. 434 pp.
A lucid historical account of physics since the Renaissance, it combines subject and chronological approaches, stressing interrelationships.

b. Buckley, H. *A Short History of Physics.* (Second Edition.) London: Methuen and Company, 1929. 263 pp.
This interesting presentation is constructed upon the framework of successive physical theories, such as planetary, atomic, kinetic and quantum.

c. Cajori, Florian. *A History of Physics.* (Revised Edition.) New York: The Macmillan Company, 1929. 424 pp.
Chronologically arranged throughout, it prints physicists' names in bold face type. The final chapter (pp. 387-406) traces the evolution of various physical laboratories.

Among noteworthy German references are:

a. Ramsauer, Carl. *Grundversuche der Physik in Historischer Darstellung.* Vol. 1. Berlin: Springer, 1953. 189 pp.

b. Hoppe, Edmund. *Geschichte der Physik.* Braunschweig: Friedrich Vieweg und Sohn, 1926. 536 pp.

c. Lange, Heinrich. *Geschichte der Grundlagen der Physik.* Freiburg/München: Verlag Karl Alber, 1954-1961. 2 vols.

Science histories.

Histories of physics *per se* may be supplemented with more general chronicles, which have the advantage of showing the relationship with science as a whole. Some well-written single volume works are:

a. Dampier-Whetham, William C. *A History of Science and its Relations with Philosophy and Religion.* (Fourth Edition.) Cambridge, England: At the University Press, 1949. 527 pp.

b. Jeans, Sir James H., and Grant, P. J. *The Growth of Physical*

Science. (Second Edition.) Cambridge, England: At the University Press, 1951. 364 pp.
The chapter headings of this interesting work are: Remote beginnings; Ionia and early Greece; Science in Alexandria; Science in the dark ages; Birth of modern science; Century of genius; The two centuries after Newton; The era of modern physics.

c. Taylor, F. Sherwood. *A Short History of Science and Scientific Thought.* New York: W. W. Norton and Company, 1949. 368 pp.
"Readings from the great scientists from the Babylonians to Einstein" appear at ends of chapters.

d. Pledge, Humphry T. *Science since 1500; A Short History of Mathematics, Physics, Chemistry, Biology.* London: His Majesty's Stationery Office, 1939. 357 pp.
This coordinates work of leading scientists, and has interesting charts and maps on the accuracy of measurements, the connection of master and pupil, the tracks of science, and the birthplaces of scientists. There is a bibliographical note (pp. 326-329) on the literature of the history of science.

e. Sedgwick, William T.; Tyler, H. W.; and Bigelow, R. P. *A Short History of Science.* (Revised Edition.) New York: The Macmillan Company, 1939. 512 pp.
Its format and typography are very attractive. An interesting chronological table (pp. 472-486) lists important names and events in the history of science and civilization in parallel columns: "Science" vs. "General History, Literature, Art, etc." A bibliography of reference books on the history of science is appended, pp. 487-500.

The early period from the fifth through nineteenth centuries is well spanned by the following books that stress scientific attitude:

a. Crombie, Alistaire C. *Augustine to Galileo.* (Second Edition.) Cambridge, Mass.: Harvard University Press, 1961. 2 vols.
Periods covered by volumes: 1, Middle ages, 5th-13th centuries; and 2, Later middle ages and early modern, 13th-17th centuries.

b. Butterfield, Herbert. *The Origins of Modern Science, 1300-1800.* (New Edition.) London: G. Bell and Sons, 1957. 242 pp.

c. Hall, Alfred R. *The Scientific Revolution, 1500-1800.* (Second Edition.) London: Longmans, 1962. 394 pp.

d. Singer, Charles J. *A Short History of Scientific Ideas to 1900.* Oxford: At the Clarendon Press, 1959. 525 pp.
(Based on his earlier *Short History of Science.*)

Lecture series are often permanently recorded in print. There is no more interesting means of communicating historical information than through sponsored lecture series, especially when the lecturers are noted scientists familiar with all phases of their subject, and gifted with powers of lucid exposition. As noteworthy examples, three such programs may be mentioned:

a. *Background to Modern Science.* Ten lectures at Cambridge arranged by the History of Science Committee 1936, by F. M. Cornford, Sir W. Dampier, Lord Rutherford, W. L. Bragg, F. W. Aston, Sir A. S. Eddington, J. A. Ryle, G. H. F. Nuttall, R. C. Punnett, J. B. S. Haldane. Edited by Joseph Needham and Walter Pagel. New York: The Macmillan Company, 1940. 243 pp.
The roster includes distinguished participants in scientific advance, 1895 to 1935. For example, Rutherford's "Forty Years of Physics" (pp. 49-74) sketches important work in radioactivity and atomic structure; Aston outlines "Forty Years of Atomic Theory" (pp. 93-114).

b. *Turning Points in Physics.* A series of lectures given at Oxford University. New York: Interscience Publishers, 1959. 192 pp.
Concerning this, Lindsay states:

> The history of science like history in general can be approached from many points of view. One of the most fascinating is that which sees in certain epochs decisive changes in the direction of theory and the introduction of radically new concepts. This is the theme of the present volume.[5]

c. *Science in Progress,* 15th Series, edited by Wallace R. Brode. New Haven, Conn.: Yale University Press, 1966. 417 pp.
In 1936 Sigma Xi conceived of this National Lectures series to be sponsored by the Society.

For more comprehensive treatment than provided by the various single volumes of standard length so far enumerated under Physics Histories and Science Histories, one turns to works of the master historians, notably the late George Sarton. His most impressive set is:

Sarton, George. *Introduction to the History of Science.* Published for the Carnegie Institution of Washington. Baltimore: Williams and Wilkins Company, 1927-1948. 3 vols. in 5.
This awe-inspiring treatise, covering meticulously the unfolding of science from Homer's time, is characterized by Sarton as an overall guide:

---

5. R. B. Lindsay, in review. *Physics Today,* 12 (No. 12): 54-55, December 1959.

> Critics who like to dispose of another man's work with a single label, should not call mine a dictionary or a bibliography but rather, if they please, a map—a scientific map with full indication of the sources.[6]

The author further states that he emphasizes interrelationships, and provides a framework rather than the whole account. The work is arranged in half-century periods; then by subject field. It is not intended to be read through. Sarton suggests, instead, reading the survey in the introductory chapter and the first chapter of each book, and consulting other sections when necessary for details. Coverage is: (Vol. 1) From 9th century B.C. through 11th century; (Vol. 2, Pt. 1) 12th century; (Vol. 2, Pt. 2) 13th century; (Vol. 3) 14th century.

The foregoing should not be confused with the author's initial volumes of a new series, which Harvard University Press is making plans to continue:

Sarton, George. *A History of Science.* Cambridge, Mass.: Harvard University Press, 1952-1959. 2 vols.

Vol. 1: Ancient science through the golden age of Greece; Vol. 2: Hellenistic science and culture in the last three centuries B.C. (The second volume was published posthumously.)

Sarton good-naturedly explained why he had undertaken still another multi-volume set:

> Many years ago, soon after the publication of volume 1 of my *Introduction,* I met one of my old students as I was crossing the Yard, and invited him to have a cup of coffee with me in a cafeteria of Harvard Square. After some hesitation, he told me, "I bought a copy of your *Introduction* and never was so disappointed in my life. I remembered your lectures, which were vivid and colorful, and I hoped to find reflections of them in your big volume, but instead I found nothing but dry statements, which discouraged me." I tried to explain to him the purpose of my *Introduction,* which was severe and uncompromising; a great part of it was not meant to be read at all but to be consulted, and I finally said, "I may be able perhaps to write a book that pleases you more."
>
> Ever since, I have often been thinking of this book which reproduces not the letter but the spirit of my lectures.[7]

Other comprehensive works include:

a. Wolf, A. *A History of Science, Technology, and Philosophy in the 16th and 17th Centuries.*[8] New York: The Macmillan Company, 1935. 692 pp.;

---

6. *Op. cit.*, Vol. 2, Pt. 1, p. vii.
7. *Op. cit.*, Vol. 1, p. vii.
8. New editions, with some corrections and later references, have been

. . . *in the Eighteenth Century.*[8] New York: The Macmillan Company, 1939. 814 pp.
This superbly illustrated set is arranged by fields of science, and has sections on scientific instruments of the times. Wolf champions the breadth of outlook recognized as essential by modern educators in stating:

> An encyclopaedic enterprise like the present may appear to be an anachronism in an age of extreme specialization. It is widely recognized, however, that the tendency toward a narrow specialism has already gone too far. The contemporary close relationship of science and philosophy, and the growing interest in the history and development of science, may be regarded as evidence of a growing recognition of the need of a wider outlook.[9]

b. Thorndike, Lynn. *History of Magic and Experimental Science.* New York: Columbia University Press, 1923-1958. 8 vols.
The inclusion of "magic" in this well-documented treatise is explained by Thorndike as follows:

> My idea is that magic and experimental science have been connected in their development; that magicians were perhaps the first to experiment; and that the history of both magic and experimental science can be better understood by studying them together.[10]

Coverage is as follows: (Vols. 1-2) First thirteen centuries; (Vols. 3-4) 14th and 15th centuries; (Vols. 5-6) 16th century; and (Vols. 7-8) 17th century.

c. Taton, René, editor. *History of Science.* New York: Basic Books, Inc., 1963-1966. 4 vols.
Volume coverage is: 1, Ancient and medieval to 1450; 2, 1450-1800; 3, Nineteenth century; and 4, Twentieth century.

d. Singer, Charles J., *et al.*, editors. *A History of Technology.* Oxford: At the Clarendon Press, 1954-1958. 5 vols.
A monumental set depicting the role of science and technology in the development of civilization.

The development of scientific ideas is traced in:

a. Whewell, William. *History of the Inductive Sciences from the Earliest to the Present Time.* (Third Edition.) New York: D. Appleton and Company, 1858. 2 vols.
As Macfarlane states:

---

prepared by Douglas McKie. (London: George Allen and Unwin, Ltd., 1950-1952.)

9. *Op. cit.,* [Vol. 1], p. xxv.
10. *Op. cit.,* Vol. 1, p. 2.

> In 1837 Whewell finished the first part of his *History of the Inductive Sciences.* In this book he notes the *epochs* when the great steps were made in the principal sciences, the *preludes* and the *sequels* of these epochs, and the way in which each step was essential to the next. He attempts to show that in all great inductive steps the type of the process has been the same. The prominent facts of each science are well selected and the whole is written with a vigor of language and a facility of illustration rare in the treatment of scientific subjects.[11]

b. Wiener, Philip P., and Noland, Aaron, editors. *Roots of Scientific Thought; A Cultural Perspective.* New York: Basic Books, Inc., 1957. 677 pp.
Articles from the *Journal of the History of Ideas* in four groups: Classical heritage; Rationalism to experimentalism; Scientific revolution; and World-machine to cosmic evolution.

c. Gillispie, Charles C. *The Edge of Objectivity; An Essay in the History of Scientific Ideas.* Princeton, N. J.: Princeton University Press, 1960. 562 pp.
The author states (p. 521) that the

> . . . purpose is to set out in narrative form what I take to be the structure in the history of classical science. This I find in the route which the advancing edge of objectivity has in fact taken through the study of nature from one science to another.

Special histories.

It will be helpful to indicate some of the areas of physics for which specialized historical accounts are available:

Atomism

a. Gregory, Joshua C. *A Short History of Atomism; from Democritus to Bohr.* London: A. and C. Black, Ltd., 1931. 258 pp.
On pp. 249-252 appears a chronological summary of important events in the history of atomism since 420 B.C.

b. Melsen, Andrew G. M. van. *From Atomos to Atom: The History of the Concept Atom.* New York: Harper and Row, 1960. 240 pp.
This Harper Torchbook is a reprint of the 1952 translation (Duquesne University Press).

Elasticity

a. Todhunter, Isaac. *A History of the Theory of Elasticity and of the Strength of Materials from Galilei to the Present Time,* edited and

---

11. A. Macfarlane, *Lectures on Ten British Physicists of the Nineteenth Century,* p. 89. New York: John Wiley and Sons, 1919.

completed by Karl Pearson. Cambridge, England: At the University Press, 1886-1893. 2 vols. in 3.

b. Timoshenko, Stephen. *History of Strength of Materials with a Brief Account of the History of Theory of Elasticity and Theory of Structures.* New York: McGraw-Hill Book Company, 1953. 452 pp.

Electricity; Magnetism

This field boasts more histories than any other. A chronological listing is provided by:

National Electrical Manufacturers Association. *A Chronological History of Electrical Development from 600 B.C.* New York: The Association, 1946. 106 pp.

Well-documented accounts of the development of electrical theory are:

a. Whittaker, Sir Edmund T. *A History of the Theories of Aether and Electricity.* New York: Philosophical Library, 1951-1954. 2 vols.

Coverage is as follows: (Vol. 1) The classical theories; (Vol. 2) The modern theories, 1900-1926; a third volume is contemplated.

b. Benjamin, Park. *A History of Electricity (The Intellectual Rise in Electricity) from Antiquity to the Days of Benjamin Franklin.* New York: John Wiley and Sons, 1898. 611 pp.

c. Daujat, Jean. *Origines et Formation de la Théorie des Phénomènes Électriques et Magnétiques.* Paris: Hermann et Cie., 1945. 3 vols.

Volume coverage is: (Vol. 1) Antiquity and Middle Ages; (Vol. 2) 17th Century; and (Vol. 3) 18th Century.

See also Mottelay's great bibliographical history.

Shorter historical narratives include:

a. Miller, Dayton C. *Sparks, Lightning, Cosmic Rays; An Anecdotal History of Electricity.* New York: The Macmillan Company, 1939. 192 pp.

This embodies lecture material and demonstration apparatus used at the Christmas week lectures for young people at the Franklin Institute. (*Cf.* the similar series at the Royal Institution.)

b. Roller, Duane, and Roller, D. H. D. *The Development of the Concept of Electric Charge; Electricity from the Greeks to Coulomb.* Cambridge, Mass.: Harvard University Press, 1954. 97 pp. (Harvard Case Studies.)

c. Greenwood, Ernest. *Amber to Amperes; The Story of Electricity.* New York: Harper and Brothers, 1931. 332 pp.

d. Still, Alfred. *Soul of Amber; The Background of Electrical Science.* New York: Murray Hill Books, Inc., 1944. 274 pp.

e. Still, Alfred. *Soul of Lodestone; The Background of Magnetical Science.* New York: Murray Hill Books, Inc., 1946. 233 pp.

See also "Resource Letter ECAN-1 on the Electronic Charge and Avogadro's Number," by David L. Anderson. *American Journal of Physics,* 34: 2-8, January 1966.

Heat; Energy

An interesting "Sketch for a History of Early Thermodynamics" and another ". . . of the Kinetic Theory of Gases" appear in *Physics Today,* 14 (No. 2): 32-42, February 1961; and 14 (No. 3): 36-39, March 1961; respectively. By E. Mendoza.

Books include:

a. Roller, Duane. *The Early Development of the Concepts of Temperature and Heat; The Rise and Decline of the Caloric Theory.* Cambridge, Mass.: Harvard University Press, 1950. 106 pp. (Harvard Case Studies.)

b. Gregory, Joshua C. *Combustion from Heracleitos to Lavoisier.* London: Edward Arnold and Company, 1934. 231 pp.

c. Ellis, Oliver C. de C. *A History of Fire and Flame.* London: The Poetry Lovers' Fellowship, 1932. 436 pp.

An unusual book, but "the work of one who is both scientist and poet; and thus it includes in its rich and vivid discussion the twofold majesty of Fire—its majesty in material events, and its majesty in the thought of man." [12]

d. Mott-Smith, Morton C. *The Story of Energy.* New York: D. Appleton-Century Company, 1934. 306 pp.

Its purpose (p. vii) is "to trace the story of man's conquest of energy, to describe the scientific discoveries that made it possible, and the chief ways that energy is applied to useful purposes." Steam engine cycles, theories of heat, conservation of energy, entropy, etc., are treated.

e. Thirring, Hans. *Energy for Man: Windmills to Nuclear Power.* (Second Edition.) Bloomington, Ind.: Indiana University Press, 1958. 409 pp.

A well-documented narrative of man's use of energy resources throughout time. This has also been reprinted as a Harper Torchbook (1962).

---

12. *Op. cit.,* p. xi.

f. Wilson, Mitchell A. *Energy.* New York: Time, Inc., 1963. 200 pp. (Life Science Library.)
A colorful presentation of energy sources and use.

See also books on energy by B. Chalmers and by D. W. Theobald, via author index.

See also "Resource Letter EEC-1 on the Evolution of Energy Concepts from Galileo to Helmholtz," by Theodore M. Brown. *American Journal of Physics,* 33: 759-765, October 1965.

Mechanics

The outstanding classic is:

Mach, Ernst. *The Science of Mechanics; A Critical and Historical Account of its Development.* (Sixth English Edition, with revisions through the ninth German edition.) La Salle, Ill.: The Open Court Publishing Company, 1960. 634 pp.
Noted as a lucid developmental account conducive to more complete understanding of the principles of mechanics, it includes a chronological table of important early works, pp. 616-618.

Other summaries are:

a. Girvin, Harvey F. *A Historical Appraisal of Mechanics.* Scranton, Pa.: International Textbook Company, 1948. 275 pp.

b. Ray, David H. *A History of Mechanics.* Lancaster, Pa.: The Author, 1910. 147 pp.
There is a chronological table on pp. 134-135 listing important contributions.

c. Dugas, René. *A History of Mechanics.* Translated into English by J. R. Maddox. Neuchâtel, Switzerland: Éditions du Griffon; New York: Central Book Company, 1955. 671 pp.

d. Dugas, René. *Mechanics in the Seventeenth Century.* Translated into English by Freda Jacquot. Neuchâtel, Switzerland: Éditions du Griffon; New York: Central Book Company, 1958. 612 pp.

Evolution of fundamental concepts is traced in:

a. Jammer, Max. *Concepts of Force; A Study in the Foundations of Dynamics.* Cambridge, Mass.: Harvard University Press, 1957. 269 pp.

b. Hesse, Mary B. *Forces and Fields; The Concept of Action at a Distance in the History of Physics.* Edinburgh: Thomas Nelson and Sons, 1961. 318 pp.
This traces from ancient to modern times philosophical, mechanical and electrical ideas on interaction.

c. Jammer, Max. *Concepts of Mass, in Classical and Modern*

Physics. Cambridge, Mass.: Harvard University Press, 1961. 230 pp.

d. Jammer, Max. *Concepts of Space; The History of Theories of Space in Physics.* Cambridge, Mass.: Harvard University Press, 1954. 196 pp.

See also Jammer's book under Quantum theory.

Optics; Color

a. Hoppe, Edmund. *Geschichte der Optik.* Leipzig: J. J. Weber, 1926. 263 pp.
This covers from 5000 B.C. to the period of the quantum theory, and has an extensive bibliography, pp. 236-251.

b. Mach, Ernst. *The Principles of Physical Optics; An Historical and Philosophical Treatment.* London: Methuen and Company, 1926. 324 pp.
The author states (p. vii):

> I have endeavoured to show, from a critical and psychological standpoint, how the ideas concerning the nature of light have been moulded at the hands of prominent individual workers, what transformations these ideas have had to undergo on account of the revelation of new facts and by reason of the views associated with them, and how the general concepts of optics develop from these.

c. Halbertsma, K. T. A. *A History of the Theory of Colour.* Amsterdam: Swetz and Zeitlinger, 1949. 267 pp.

d. Birren, Faber. *Color; A Survey in Words and Pictures, from Ancient Mysticism to Modern Science.* New Hyde Park, N. Y.: University Books, 1963. 223 pp.

e. Harvey, E. Newton. *A History of Luminescence; From the Earliest Times Until 1900.* Philadelphia: American Philosophical Society, 1957. 692 pp. (Its *Memoirs,* Vol. 44.)

Sound

Miller, Dayton C. *Anecdotal History of the Science of Sound.* New York: The Macmillan Company, 1935. 114 pp.

Current events.

To keep abreast of current scientific developments one scans the regular and science sections of the *New York Times* and other newspapers, a chronological key to all being furnished by the *New York Times Index;* and reads the appropriate periodicals, whose contents are indexed as described in Chapter II. These range from the popular *Science News, Scientific American,* and *Popular Science Monthly,* to the standard journals, *Science,* and *Nature.* More specialized are *Journal of Applied Physics, Physics Today, Optical Society of America*

Journal, and countless others. For the history of science there is the international review called *Isis,* and also the British *Annals of Science.* General review serials (like *Reviews of Modern Physics*), and specialized media (such as *Advances in Electronics . . .*) have previously been cited. Shorter yearly reviews of physics developments may be found in the annual supplementary volumes to the *McGraw-Hill Encyclopedia of Science and Technology,* the *Encyclopedia Americana* and other encyclopedias, as well as in journal résumés. Overview articles by distinguished scientists are appended to the *Annual Reports* of the Smithsonian Institution, fifty of which are represented in:

Smithsonian Institution. *Smithsonian Treasury of Science,* edited by Webster P. True. New York: Simon and Schuster, Inc., 1960. 3 vols.

In the sequel, twenty-six scientific frontiers are explored by specialists:

Smithsonian Treasury of 20th Century Science, edited by Webster P. True. New York: Simon and Schuster, Inc., 1966. 544 pp.

See also the following account:

Karp, Walter. *The Smithsonian Institution.* Washington, D. C.: The Institution, 1965. 125 pp.

For centennial and other major anniversaries (birth or death) of famous physicists, see E. Scott Barr's annual articles in *American Journal of Physics* beginning 1957. Brief accounts of lives and contributions are included.

Summary

Full appreciation of physics as a cultural subject requires historical perspective. Background literature has been discussed, ranging from relatively short narratives of physics or general science to monumental treatises. Many special fields have their own historical accounts, often overlooked among the larger body of technical material which they nevertheless clothe with reality. Finally, the record is brought to date by mentioning sources of current events, from which will be selected the items of historical tomes yet to be written.

CHAPTER IV

BIOGRAPHICAL APPROACH

Biography and Source Material

This chapter reviews sources of biographical data, both current and retrospective. Because personal writings reveal human traits and scientific outlook beyond mere subject content, collected works and source extracts have been placed after the strictly biographical material.

Overview

The introduction to the previous chapter applies equally well to the present one. Individuals, of course, are the makers of science history, which is the record of their achievements among their fellows against world backgrounds. Hence historical narratives are rich in biographical detail, and biographies shed interesting sidelights on history. Collected writings and excerpts fulfill a similar function:

> . . . It is doubtful if any description brings out the scientific outlook of a period, or the experimental handicaps under which the foundations of the subject were laid, as forcibly as do the original papers. Furthermore the experimental difficulties, precautions, corrections, and incidental technique which have been found necessary for the attainment of extreme accuracy are nowhere so clearly shown as in the description of the investigators themselves.[1]

Collective Biography

In a general or preliminary search for relatively brief biographical data, one turns first to the books that deal with more than one scientist.

Encyclopedias.

The most comprehensive biographical undertaking for scientists of all countries and eras is:

Poggendorff, Johann C. *Biographisch-Literarisches Handwörterbuch der Exakten Naturwissenschaften.* Leipzig: J. A. Barth, 1863-

1. G. P. Harnwell and J. J. Livingood, *Experimental Atomic Physics,* p. ix. New York: McGraw-Hill Book Company, 1933.

1904; Berlin: Verlag Chemie, 1925-1940; Akademie Verlag, 1955-7 vols. in 15 (through Vol. 7A). (In Process.)
Part A of the seventh volume covers Germany, Austria and Switzerland (1932-1953); other countries to be in Part B. This important reference work furnishes biographical details and writings.

Leading scientists from ancient to modern times appear in:

a. Howard, A. V. *Chambers's Dictionary of Scientists.* New York: E. P. Dutton and Company, 1951. 499 pp.

b. Asimov, Isaac. *Asimov's Biographical Encyclopedia of Science and Technology.* Garden City, N. Y.: Doubleday and Company, 1964. 662 pp.
Biographies are arranged chronologically, giving continuity.

c. Ireland, Norma O. *Index to Scientists of the World, from Ancient to Modern Times: Biographies and Portraits.* Boston, Mass.: F. W. Faxon Company, 1962. 662 pp.
In 338 collective sources, specific location is given of biographies, portraits and contributions.

See also index under Portraits.

For present-day scientists of the world there is:

McGraw-Hill Modern Men of Science: 426 Leading Contemporary Scientists, presented by the editors of the *McGraw-Hill Encyclopedia of Science and Technology.* New York: McGraw-Hill Book Company, 1966. 620 pp.
Achievements are described for major award recipients since 1940.

Other comprehensive sources, such as Webster's *Biographical Dictionary* (1962) and *Biographie Universelle* (Michaud) are listed under Biography in Winchell. Here one finds also the various national encyclopedias, such as the British *Dictionary of National Biography*, and the *Dictionary of American Biography*, both of which exclude living persons. On the other hand, the *National Cyclopaedia of American Biography* includes contemporaries, thus supplementing the directories mentioned in the next paragraph. The *Encyclopedia Americana* and others should not be overlooked as sources of biographical articles on famous scientists of the past.

Directories.

For present day scientists of the United States and Canada, there is a very extensive compilation of brief biographical accounts, *viz.*:

American Men of Science: A Biographical Directory, edited by the Jaques Catell Press. *The Physical and Biological Sciences.* (Eleventh

Edition.) New York: R. R. Bowker Company, 1965-1967. 6 vols.; plus 4 supplements cumulated into 2.
Lists 130,000 scientists, representing a 30% increase over the previous edition. The supplements update sections of the alphabet. (Volumes for the social and behavioral sciences will follow.)

For American and Canadian scientists there is also:

Leaders in American Science, edited by Robert C. Cook. (Seventh Edition: 1966-1967.) Nashville, Tenn.: Who's Who in American Education, Inc., 1967. 681 pp.

Brief biographies and chief publications of British scientists appear in:

Directory of British Scientists 1966-1967. (Third Edition.) New York: R. R. Bowker Company, 1966. 2 vols.

For West European scientists there is a new multi-volume directory published by Francis Hodgson Ltd. entitled "*Who's Who in Science in Europe.*"

Russian scientists appear in:

a. Turkevich, John. *Soviet Men of Science: Academicians and Corresponding Members of the Academy of Sciences of the U.S.S.R.* Princeton, N. J.: D. Van Nostrand Company, 1963. 441 pp.

b. *Who's Who in Soviet Science and Technology,* compiled by Ina Telberg. (Second Edition.) New York: Telberg Book Company, 1964. 301 pp.

For atomic scientists see:

Who's Who in Atoms; An International Reference Book. (Fourth Edition.) London: Harrap Research Publications, 1965. 1456 pp.

Longer résumés may be found in the more general *Who's Who in America;* its British counterpart, *Who's Who;* and sometimes in *Who's Who in Engineering* if the individual is a technologist as well. *Current Biography,* an H. W. Wilson Company monthly and annual publication, provides lengthy informal sketches of scientists and others in the news. *Biographical Memoirs,* published by the National Academy of Sciences, and the Royal Society of London's *Biographical Memoirs of Fellows,*[2] both present substantial biographies of recently deceased scientists, with their portraits and lists of writings. At intervals the American Physical Society, the Institute of Physics and the Physical Society, the American Society for Engineering Education,

---

2. Vol. 1—1955—in continuation of its *Obituary Notices,* Vols. 1-9, 1932-1954.

and other groups publish identity lists of their members as special bulletins.

Nobel prize winners may be found in:

a. *Nobel Lectures, Including Presentation Speeches and Laureates' Biographies. Physics:* 1901-1962. Edited by the Nobel Foundation. Amsterdam, New York, etc.: Elsevier Publishing Company, 1964-1967. 3 vols.
Laureate lectures are made completely available in English, whereas they appear in the language of presentation in the official annual, *Les Prix Nobel.* The series continues for Chemistry and other fields.

b. Heathcote, Niels H. de V. *Nobel Prize Winners in Physics 1901-1950.* New York: Henry Schuman, 1953. 473 pp.
This is still useful for work summaries and evaluative comments, furnished in addition to brief biographies and speech extracts.

c. *Nobel, The Man and His Prizes,* edited by the Nobel Foundation. (Second Edition.) Amsterdam, New York, etc.: Elsevier Publishing Company, 1962. 690 pp.; 3 tables.

See also the Nobel laureates' portraits in Weber, Manning and White's *College Physics.*

For record of medals, prizes, etc., presented by North American science groups, see:

Special Libraries Association. Science-Technology Division. *Handbook of Scientific and Technical Awards in the United States and Canada, 1900-1952,* edited by Margaret A. Firth. New York: The Association, 1956. 491 pp.
Awards are described and recipients listed, with literature references for presentations since 1929.

See also mention of various awards in the personal notes section of *Physics Today.*

Narrative accounts.

Unified running treatments in narrative form might possibly belong in the "Historical Approach," but the following are more distinctly biographical in content and title:

a. Hart, Ivor B. *The Great Physicists.* (Second Edition.) London: Methuen and Company, 1934. 138 pp.
This has such chapter heads as: The physicists of classical antiquity; The dawn of experimental physics; Newtonian physics.

b. Fraser, Charles G. *Half-Hours with Great Scientists; The Story of Physics.* New York: Reinhold Publishing Corporation, 1948. 527 pp.

The "stories" of mechanics, acoustics, optics, etc., are recounted in very interesting fashion, and source material is included.

c. Gamow, George. *Biography of Physics.* New York: Harper and Brothers, 1961. 338 pp.
Physics interest is awakened by observing leaders of their eras.

d. Hart, Ivor B. *Makers of Science; Mathematics, Physics and Astronomy.* London: Oxford University Press, 1923. 320 pp.

e. Turner, D. M. *Makers of Science; Electricity and Magnetism.* London: Oxford University Press, 1927. 184 pp.
As in the other books of this group, biographical details have been skilfully woven with the development of science.

Multiple biographies.

Good examples of compilations of longer treatments of several scientists are:

a. Crowther, J. G. *Men of Science; Humphry Davy, Michael Faraday, James Prescott Joule, William Thomson, and James Clerk Maxwell.* New York: W. W. Norton and Company, 1936. 332 pp.

b. Lenard, Philipp. *Great Men of Science; A History of Scientific Progress,* translated from the Second German Edition. London: G. Bell and Sons, 1933. 389 pp.
Depicting scientists as living people, Lenard states (p. xiii):

> My joy was great to find that these personalities so well matched the greatness of their achievements, that they were fit to serve as examples to future generations both from the point of view of their work and from that of their lives.

Some of them are: Archimedes, Galileo, Pascal, Descartes, Boyle, Huygens, Newton, Cavendish, Coulomb, Volta, Davy, Young, Ohm, Faraday, and Maxwell.

c. Bolton, Sarah K. *Famous Men of Science.* (Revised Edition.) New York: Thomas Y. Crowell Company, 1938. 376 pp.
Humanized treatments show how Galileo, Newton, Faraday, Lord Kelvin, and the Curies overcame obstacles.

d. Appleyard, Rollo. *Pioneers of Electrical Communication.* London: The Macmillan Company, 1930. 347 pp.
The ten men selected are: Maxwell, Ampère, Volta, Wheatstone, Hertz, Oersted, Ohm, Heaviside, Chappe, and Ronalds.

e. Jones, Bessie J. Z., editor. *The Golden Age of Science; Thirty Portraits of the Giants of 19th-Century Science, by Their Scientific Contemporaries.* New York: Simon and Schuster, Inc., 1966. 659 pp.

Published in coöperation with the Smithsonian Institution. Some of the "giants": Darwin, Faraday, Volta, Henry, Bunsen, Oersted.

Individual Biography

When more extensive information on the life and achievements of a man of science than can be provided by general compendia is sought, one reads the longer single-subject accounts, assuming their availability.

Higgins has compiled exhaustive lists of full-length biographies of physicists and others, which readily indicate whether substantial life treatments have been published. These lists may be consulted in original form among the pages of the *American Journal of Physics*,[3] or more compactly in:

Higgins, Thomas J. *Biographies of Engineers and Scientists.* Chicago: Illinois Institute of Technology, 1949. 62 pp. (Its *Research Publications*, Vol. 7, No. 1.)

An earlier miscellaneous list was:

Pittsburgh. Carnegie Library. *Men of Science and Industry.* Pittsburgh: The Library, 1915. 189 pp.
This furnished a key to the biographical material on scientists, engineers, inventors, and physicians, in the library collection at the time.

A general current listing of biographical books and periodical articles, the *Biography Index*,[4] is frequently useful because it has a profession index in addition to its main alphabet of personal names. It appears quarterly, with periodic cumulations, and should not be confused with the same publisher's *Current Biography*, previously described under Directories. The biography sections of the *Standard Catalog* series (*for Public Libraries* and *for High School Libraries*, respectively) list recommended biographies, including some for science. In one's own library, such material is to be found in the card catalog under personal name and under the headings "Physics—Biography," "Science—Biography," and possibly "Scientists," "Physicists," etc. New publications may also be found in the *Cumulative Book Index.*

Only a few examples of individual (as opposed to multiple) biography need be cited:

---

3. T. J. Higgins, "Book-Length Biographies of Physicists and Astronomers." *American Journal of Physics*, 12: 31-39, February 1944. Also, for addenda, *ibid.*, 12: 234-236, August 1944; and 16: 180-182, March 1948.
4. Published by the H. W. Wilson Company.

a. Sullivan, John W. N. *Isaac Newton, 1642-1727.* New York: The Macmillan Company, 1938. 275 pp.

b. More, Louis T. *The Life and Works of the Honourable Robert Boyle.* London: Oxford University Press, 1944. 313 pp.

c. Milne, Edward Arthur. *Sir James Jeans; A Biography.* Cambridge, England: At the University Press, 1952. 175 pp.

d. Garbedian, Haig G. *Albert Einstein; Maker of Universes.* New York: Funk and Wagnalls Company, 1939. 328 pp.

e. Rozental, Stefan, editor. *Niels Bohr: His Life and Work as Seen by His Friends.* Amsterdam: North-Holland Publishing Company, 1967. 355 pp.

f. Kilmister, C. W. *Sir Arthur Eddington.* Oxford: Pergamon Press, 1966. 279 pp.

Occasionally a scientist will venture an autobiography:

a. Millikan, Robert A. *Autobiography.* New York: Prentice-Hall, Inc., 1950. 311 pp.

b. Thomson, Sir Joseph J. *Recollections and Reflections.* New York: The Macmillan Company, 1937. 451 pp.

A personal record of different type, focused upon experimental occurrences, is an outstanding classic:

Faraday's Diary, being the Famous Philosophical Notes of Experimental Investigation made by Michael Faraday during the years 1820-1862, edited by Thomas Martin. London: G. Bell and Sons, 1932-1936. 7 vols. and index.

Sir William H. Bragg states in the foreword (p. v.):

> He was in the habit of describing each experiment, in full and careful detail, on the day on which it was made. . . . The main interest of the Diary lies, however, quite outside the range of propositions and experimental proofs. It centres round the methods of Faraday's attack, both in thought and in experiment: it depends on the records of the workings of his mind as he mastered each research in turn, and on his attitude not only to his own researches but also to scientific advance in general.

Festschrift is the German term applied to a collection of essays prepared by a scientist's friends in his honor. Examples follow:

a. Frank, W., editor. *Max-Planck-Festschrift, 1958.* Berlin: Veb Deutscher Verlag der Wissenschaften, 1959. 413 pp.

b. *Jubilé de Marcel Brillouin. Mémoires Originaux offerts à Marcel Brillouin à l'occasion de son 80e Anniversaire.* Paris: Gauthier-Villars, 1935. 441 pp.

c. *Contributions to the Mechanics of Solids dedicated to Stephen Timoshenko by his Friends on the occasion of his Sixtieth Birthday*

Anniversary. New York: The Macmillan Company, 1938. 277 pp. The first paper is a biographical sketch by John M. Lessells, with a bibliography of Timoshenko's writings.

d. Pauli, W., editor. *Niels Bohr and the Development of Physics. Essays dedicated to Niels Bohr on the occasion of his Seventieth Birthday.* New York: McGraw-Hill Book Company, 1955. 195 pp.

e. Löwdin, Per-Olov, editor. *Quantum Theory of Atoms, Molecules and the Solid State: A Tribute to John C. Slater.* New York: Academic Press, 1966. 641 pp.

Individual Writings

Although not biographical, of course, the collected publications of scientists are included under the Biographical Approach because of personal connotation. By their works do we not indeed know them?

Bibliographies.

Examples of bibliographies that have been compiled to chart prolific writings follow:

a. Maire, Albert. *Bibliographie Générale des Oeuvres de Pascal.* Paris: L. Giraud-Badin, 1925-1927. 5 vols.
This includes critical and biographical data.

b. Gray, George J. *A Bibliography of the Works of Sir Isaac Newton.* (Second Edition.) Cambridge, England: Bowes and Bowes, 1907. 80 pp.

Closely related is the commentary, exemplified by:

Donnan, F. G., and Haas, Arthur, editors. *A Commentary on the Scientific Writings of J. Willard Gibbs.* New Haven, Conn.: Yale University Press, 1936. 2 vols.
This treatise constitutes a memorial supplement to Gibbs' works (Longmans, 1928. 2 vols.).

Collected works.

Library collections include many collective volumes of physicists' contributions, indexed in the card catalog under the heading "Physics —Collected Works" as well as under personal names. Poggendorff (cited early in this chapter) mentions them, and there is a representative list in a textbook [5] by Millikan, *et al.*

---

5. R. A. Millikan, D. Roller, and E. C. Watson, *Mechanics, Molecular Physics, Heat, and Sound.* Boston: Ginn and Company, 1937. Bibliog.: pp. 435-456. (Reprinted by M.I.T. Press, 1965.)

Random selections from among numerous available sets are:

a. Bridgman, Percy W. *Collected Experimental Papers.* Cambridge, Mass.: Harvard University Press, 1964. 7 vols.
The record of a half-century of elegant experimental work in high-pressure physics, by a Nobel laureate.

b. Fermi, Enrico. *Collected Papers.* Chicago: University of Chicago Press, 1962-1965. 2 vols. Vol. 1: Italy, 1921-1938; Vol. 2: United States, 1939-1954.
In this superb memorial collection are included historical and anecdotal accounts contributed by Fermi's former associates.

c. Lorentz, H. A. *Collected Papers.* The Hague, Netherlands: Martinus Nijhoff, 1935-1939. 9 vols.
This attractive set comprises papers in French, English, German and Dutch; introductory prefaces in English; and systematic and chronological bibliographies of the writings (Vol. 9, pp. 411-434).

d. Pauli, Wolfgang. *Collected Scientific Papers,* edited by R. Kronig and V. F. Weisskopf. New York: Interscience Publishers, 1964. 2 vols.

e. Rayleigh, John W. S. *Scientific Papers, by John William Strutt,*[6] *3d Baron Rayleigh.* Cambridge, England: At the University Press, 1899-1920. 6 vols.

f. Rutherford, Ernest Rutherford. *The Collected Papers of Lord Rutherford of Nelson.* London: George Allen and Unwin, Ltd., 1962- 4 vols. (In process.)

Source materials.

No more fascinating approach to scientific progress through the ages can be taken than through the original papers of the great men of science, revealing obstacles surmounted and techniques developed. As Silvanus P. Thompson asserts emphatically in his introduction to Mottelay's great bibliography:

> The art of scientific discovery—for it is an art—can be attained in but one way, the way of attainment in all arts, namely, by practising it. In the practice of art, the aspirant may at least learn something that all the textbooks cannot drill out of him, and which will help him in his practice, by the careful examination of the actual ways in which the discoveries of science, now facts of history, were actually made.

---

6. Note that library card catalogs use *see references* between name variants of Lords Rayleigh, Kelvin (William Thomson), *et al.* One must distinguish between successive Lords Rayleigh also; the *4th Baron Rayleigh* is Robert John Strutt.

> But, to do this, he must throw overboard for a time the systematic textbooks, he must abandon the logical expositions which embody, at second hand, or at third hand, the antecedent discoveries, and he must go to the original sources, the writings and records of the discoverers themselves, and learn from them how they set to work. The modern compendious handbooks, in which the results of hundreds of workers have been boiled down, as it were, to a uniform consistency, is exactly the intellectual pabulum which he must eschew.

This recourse to science in its nascent state is rendered delightful as well as instructive because invariably the great experimenters are also masters of lucid scientific exposition.

The standard source book is:

Magie, William F. *A Source Book in Physics.* New York: McGraw-Hill Book Company, 1935. 620 pp.
Selected passages are given (in English throughout), together with brief biographical sketches of important physicists of the period 1600-1900. This book should always be conveniently at hand for frequent use.

Other useful books having source material are:

a. Cohen, Morris R., and Drabkin, I. E. *A Source Book in Greek Science.* (New Edition.) Cambridge, Mass.: Harvard University Press, 1958. 581 pp.
Of physics interest are excerpts from Aristotle, Euclid, Archimedes, Ptolemy, Lucretius, Hero, *et al.*

b. Dampier-Whetham, William C., and Dampier-Whetham, Margaret. *Cambridge Readings in the Literature of Science.* Cambridge, England: At the University Press, 1924. 275 pp.
The extracts are arranged to outline cosmogony, atomic theories, and evolution. Archimedes, Copernicus, Galileo, Newton, Laplace, Avogadro, Sir J. J. Thomson, F. W. Aston, H. G. J. Moseley, Sir Ernest Rutherford, *et al.*, are represented.

c. Knickerbocker, William S. *Classics of Modern Science (Copernicus to Pasteur).* New York: Alfred A. Knopf, 1927. 384 pp.
Good literary style is here exemplified by passages from Copernicus, Galileo, Boyle, Huygens, Newton, Franklin, Volta, Rumford, Dalton, Avogadro, Faraday, Henry, von Helmholtz, Maxwell, and other scientists.

d. Wright, Stephen, editor. *Classical Scientific Papers—Physics.* London: Mills and Boon, Ltd., 1964. 393 pp.
Facsimiles of journal articles by Rutherford, Thomson, Geiger, Compton, Wilson, Aston, *et al.*, reporting their discoveries in atomic and field theory.

e. Schwartz, George, and Bishop, Philip W., editors. *Moments of Discovery.* New York: Basic Books, Inc., 1958. 2 vols.
Actual occasions of scientific insight from writings of over eighty scientists of all eras. Volume titles are: Origins of science; Development of modern science.

f. Garrett, Alfred B. *The Flash of Genius.* Princeton, N. J.: D. Van Nostrand Company, 1963. 249 pp.
Narration, with source extracts, of fifty-one discoveries in physics and chemistry: the key experiment and the moment of devising a theory.

g. *Science: A Course of Selected Reading by Authorities.* (Second Edition.) Nottingham, Eng.: International University Society, 1958. 322 pp.
Its four sections are: The origin and meaning of science; The universe; Matter and energy; and Science and everyday life. Also titled *Classics in Science* (New York: Philosophical Library, 1960).

Further excerpts appear in F. S. Taylor's *Short History of Science,* F. H. Law's *Science in Literature,* and M. H. Shamos' *Great Experiments in Physics.*

Source material series have been issued by several publishers. Component volumes may be traced from series cards in library catalogs, from publishers' lists, or from flyleaves of individual books.

1. English language presentations may be found in:

a. *Harper's Scientific Memoirs,* edited by Joseph S. Ames. New York: American Book Company; Harper and Brothers, 1898-1902. 15 vols.
Representative titles follow:

Barker, George F. *Röntgen Rays; Memoirs by Röntgen, Stokes, and J. J. Thomson.* 1899. 76 pp.

Crew, Henry. *The Wave Theory of Light; Memoirs by Huygens, Young, and Fresnel.* 1900. 164 pp.

Mackenzie, A. Stanley. *The Laws of Gravitation; Memoirs by Newton, Bouguer, and Cavendish.* 1900. 160 pp.

A review of this fifteen-volume series is available.[7] More recently Harper and Row have issued *Harper Science Library* reprints of Wolf's history, Dampier-Whetham's readings, and other substantial books. *Harper's Torchbooks* is another interesting series.

b. *Bell's Classics of Scientific Method.* London: G. Bell and Sons, 1922-
Typical units in this series, designed to provide the layman with re-

7. *Physical Review,* 15: 192, September 1902.

productions of the great masterpieces, and to trace the scientific thought and action leading up to them, are the following:

Roberts, Michael, and Thomas, E. R. *Newton and the Origin of Colours.* 1934. 133 pp.

Wood, Alexander. *Joule and the Study of Energy.* 1925. 88 pp.

c. *Selected Readings in Physics,* edited by D. ter Haar. Oxford: Pergamon Press, 1965-
Original papers and commentary show how various subject fields developed, e.g.:

Sanders, J. H. *The Velocity of Light.* 1965. 144 pp.

Brush, Stephen G. *Kinetic Theory.* 1965-1966. 2 vols.

Tricker, R. A. R. *The Contributions of Faraday and Maxwell to Electrical Science.* 1966. 289 pp.

d. *Classics of Science,* edited by Gerald Holton. New York: Dover Publications, 1963-
Each volume of this new series will contain the articles on one development, with a commentary and historical sidelights, as in:

Romer, Alfred, editor. *The Discovery of Radioactivity and Transmutations.* 1964. 233 pp.

Livingston, M. Stanley, editor. *The Development of High-Energy Accelerators.* 1966. 317 pp.

2. German language presentations are given in:

Ostwalds Klassiker der Exakten Wissenschaften, begründet von Wilhelm Ostwald; herausgegeben von Wolfgang Ostwald. Leipzig: Akademische Verlagsgesellschaft, 1889-
No. 248 of this series appeared in 1960.

3. French language presentations appear in *Les Maîtres de la Pensée Scientifique,* a series published in Paris by Gauthier-Villars.

Facsimiles of historical scientific works are included in other source material series, such as:

a. *The Sources of Science,* edited by Harry Woolf. New York: Johnson Reprint Corporation, 1964-
The first two were:

Essayes of Natural Experiments Made in the Academie del Cimento, Englished by Richard Waller. (Facsimile of the London 1684 edition.) 1964. 160 pp.

Boyle, Robert. *Experiments and Considerations Touching Colours.* (Facsimile of the London 1664 edition.) 1964. 423 pp.

b. *Collection History of Science.* Brussels: *Éditions Culture et Civilisation,* 1966-
Examples are:

Alembert, Jean le Rond d'. *Traité de Dynamique.* (Facsimile of the Paris 1743 edition.) 1966. 222 pp.
(Four other d'Alembert works are separately offered.)

Huygens, Christiaan. *Traité de la Lumière.* (Facsimile of the Leyde 1690 edition.) 1966. 192 pp.
(His *Horologium Oscillatorium* is also available.)

The Project on the History of Recent Physics in the United States, sponsored by the American Institute of Physics, is seriously concerned with locating, preserving and studying present-day source material, for according to W. J. King (who headed the project during the first three years of its existence):

> much has been done to gather and to exploit the documentary material extant from the earlier periods of physics, notably that deriving from the work of the great earlier physicists—Galileo, Descartes, and Newton, to name a few. However, a preliminary survey by our project staff confirmed the suspicion that comparatively little has been done to ensure that comparable material—personal letters, photographs, apparatus, laboratory records, and annotated books and reprints, for example—will be preserved from our busier and more impersonal time for the future historian of physics.

As further stated by W. James King in *Physics Today,* 15 (No. 1): 44-48, January 1962, physicists are obligated to preserve the records of their contemporaries for present and future historians to interpret. For a progress report see *Physics Today,* 17 (No. 10): 84-86, October 1964.

See also index under Classics.

Summary

Familiarity with the struggles and achievements of men of science endows them with vitality and substance, qualities sorely lacking when merely names are learned in connection with important laws or discoveries. Multiple lives merge into historical accounts. Biographical sources, both current and retrospective, have been mentioned, notably Higgins' extensive lists. From individual writings, embodied in collected works or read conveniently in source book form, comes a feeling of participation in the hard original search for new physical data and concepts.

CHAPTER V

EXPERIMENTAL APPROACH

Experimentation, Equipment, and Techniques

Experimentation characterizes the work of the physicist. His laboratory and its procedures, which underlie broader research contributions, are the subject of this chapter.

Overview

When Galileo demonstrated by actual test (probably *not* publicly from the Tower of Pisa [1]) that bodies fall with the same acceleration regardless of mass, he was acting in accordance with the scientific method. Stewart has prepared an extremely concise summary [2] of the scientific method, which is also exemplified in:

Conant, James B., and Nash, L. K., editors. *Harvard Case Histories in Experimental Science.*[3] Cambridge, Mass.: Harvard University Press, 1957. 2 vols.

Further details may be found in:

a. Davies, J. T. *The Scientific Approach.* New York: Academic Press, 1965. 100 pp.

Chapter headings are: Origin of theories; Testing of theories; Confirmation and confidence; Engineering and science; Simple laws of science; Prediction and probability; and Science and society.

b. Lastrucci, Carlo L. *The Scientific Approach; Basic Principles of the Scientific Method.* Cambridge, Mass.: Schenkman Publishing Company, 1963. 257 pp.

c. Wolf, Abraham. *Essentials of Scientific Method.* London: George Allen and Unwin, Ltd., 1925. 160 pp.

d. Jeffreys, Sir Harold. *Scientific Inference.* (Second Edition.) Cambridge, England: At the University Press, 1957. 236 pp.

---

1. Cf. L. Cooper, *Aristotle, Galileo, and the Tower of Pisa.* Ithaca, N. Y.: Cornell University Press, 1935. 102 pp.; and its review by L. W. Taylor in *American Physics Teacher,* 4: 44-45, February 1936.
2. O. M. Stewart, and N. S. Gingrich, *Physics,* pp. 700-704. (Fifth Edition.) Boston: Ginn and Company, 1950.
3. See J. W. Shirley, "The Harvard Case Studies . . . : The Evolution of an Idea." *American Journal of Physics,* 19: 419-423, October 1951.

e. Tricker, R. A. R. *The Assessment of Scientific Speculation; A Survey of Certain Current Views.* New York: American Elsevier Publishing Company, 1965. 200 pp.
Four different views of the scientific method: 1, View arising out of considerations of probability; 2, Scientific theory as a logical construction out of observables; 3, View of sciences as processes of falsification of hypotheses tentatively put forward; and 4, Descriptions of nature by analogy.

f. Margenau, Henry, and Bergamini, David. *The Scientist.* New York: Time, Inc., 1964. 199 pp. (Life Science Library.)
A colorful book on the scientist and his methods.

Scientific procedures of master experimental physicists are revealed in:

Shamos, Morris H. *Great Experiments in Physics.* New York: Holt, Rinehart and Winston, 1959. 370 pp.
Following a discussion of the scientific method, twenty-four great experiments are presented with marginal comments alongside the original texts. *Cf.* T. W. Chalmers' *Historic Researches.*

Owen warns that the college laboratory does not automatically nurture scientific thinking habits:

> But how well does the ordinary laboratory experience contribute to the development of skill in applying the scientific method and of desirable attitudes and habits that should go with it? Consider that in the usual experiment someone else states the problem, develops the theory showing how general principles apply, outlines how the information is to be obtained and how it is to be interpreted and, in many cases, determines what the conclusions must be. Originality, openmindedness, even intellectual honesty are not encouraged. The student is expected to follow the reasoning outlined for him and thereby learn something about the scientific method. The assumption must be that he cannot do this for himself. If so, the experiment does not help him much in learning to apply the method to his own problems.[4]

A teacher's manual of ideas for modifying existing laboratory programs is available:

American Association of Physics Teachers. Committee on Apparatus for Educational Institutions. *Novel Experiments in Physics; A Selection of Laboratory Notes Now Used in Colleges and Universities.* New York: American Institute of Physics, 1964. 451 pp.

---

4. G. E. Owen, "Some Contributions the Physics Laboratory can make to General Education." *American Journal of Physics,* 17: 270-272, May 1949.

Experimental physics possesses an encyclopedic *Handbuch der Experimental Physik* of historic interest.

Physical Laboratory

Design.

The planning of laboratories for maximum efficiency is detailed in:

a. Coleman, H. S., editor. *Laboratory Design: National Research Council Report on Design, Construction and Equipment of Laboratories.*[5] New York: Reinhold Publishing Corporation, 1951. 393 pp.
This includes a thorough discussion of materials and equipment for teaching and industrial laboratories. Part IV presents short illustrated descriptions of various laboratories.

b. Institute of Physics and the Physical Society. *The Design of Physics Research Laboratories.* London: Chapman and Hall, Ltd., 1959. 108 pp.
This 1957 symposium was attended by architects and physicists for joint discussions.

c. Munce, James F. *Laboratory Planning.* London: Butterworths Scientific Publications, 1963. 360 pp.
British, American and German designs are compared. Some chapters: History of the laboratory; Basic concepts; Institutional laboratories; Research laboratories; Fittings and finishes; Furniture and fabric; and Safety precautions.

The proper setting is needed:

Palmer, Ralph R., and Rice, William M. *Modern Physics Buildings; Design and Function.* New York: Reinhold Publishing Corporation, 1961. 324 pp.
This is a joint project report of the American Association of Physics Teachers and the American Institute of Physics. There is a 1965 supplement, *Physics Buildings Today,* and a useful *Checklist for Physics Buildings.*

Further descriptions useful for comparative purposes are found in brochures occasionally emanating from various institutions, such as:

a. *A History of the Cavendish Laboratory, 1871-1910.* London: Longmans, Green and Company, 1910. 342 pp.
This interesting account, prepared by the associates of Sir J. J. Thom-

---

5. See also *Laboratory Design for Handling Radioactive Materials,* a 140-page discussion of layout, shielding, ventilation, etc., comprising the *Proceedings of the Third Research Correlation Conference;* available from N. R. C. Building Research Advisory Board in 1952.

son, depicts the layout, history and guiding philosophy of this famous laboratory [6] at Cambridge University, England.

b. *The Physical Laboratories of the University of Manchester; A Record of 25 Years' Work.* Manchester, England: At the University Press, 1906. 142 pp.

Another early laboratory is described with four plans.

c. *The Physical Laboratories of Harvard University.* Cambridge, Mass.: Harvard University Press, 1932. 47 pp.

d. Langdon-Davies, John. *Jubilee Book of the National Physical Laboratory.* London: H. M. Stationery Office, 1951. 103 pp.

Instruments.

Interesting accounts of old instruments associated with research and teaching at Oxford and Cambridge Universities, respectively, appear in:

a. Gunther, Robert T. *Early Science in Oxford,* Vol. I, Pt. iii, pp. 189-324. London: Oxford University Press, 1923.

b. Gunther, Robert T. *Early Science in Cambridge.* London: Oxford University Press, 1937. 513 pp.

Harvard University's collection of early scientific instruments for the study of natural philosophy is featured in:

Cohen, I. Bernard. *Some Early Tools of American Science.* Cambridge, Mass.: Harvard University Press, 1950. 201 pp.

Other historical collections appear in:

a. Bedini, Silvio A. *Early American Scientific Instruments and Their Makers.* Washington, D. C.: Smithsonian Institution, 1964. 196 pp. (Bulletin 231.)

b. Billmeir, J. A. *Scientific Instruments, 13th-19th Century.* (Second Edition.) Oxford: Museum of the History of Science, 1955-1957. 2 vols. (including the supplement).

Wolf's historical treatises also have sections on early scientific instruments.

The following compendia of modern instruments may be useful to the physicist although considerable space is devoted to those beyond his special interests:

---

6. See also: Alexander Wood, *The Cavendish Laboratory.* Cambridge, England: At the University Press, 1946. 58 pp.; E. Larson, *The Cavendish Laboratory.* New York: Franklin Watts, Inc., 1962. 95 pp.; and W. L. Bragg, "A Center of Fundamental Research." *Physics Today,* 6 (No. 1): 18-19, January 1953.

a. Cooper, Herbert J. *Scientific Instruments,* I-II. Brooklyn, N. Y.: Chemical Publishing Company, 1946-1949. 2 vols.

b. Brown, Earle B. *Optical Instruments.* Brooklyn, N. Y.: Chemical Publishing Company, 1945. 567 pp.

c. Martin, Louis C. *Optical Measuring Instruments.* London: Blackie and Son, 1924. 270 pp.

(See also his more recent *Technical Optics,* and *Geometrical Optics.*)

d. Soisson, Harold E. *Electronic Measuring Instruments.* New York: McGraw-Hill Book Company, 1961. 352 pp.

See also index under Optics, applied, and Measurements, special.

There is an instrumentation bibliography:

Guide to Instrumentation Literature. Washington, D. C.: Government Printing Office, 1965. 220 pp.

(U. S. National Bureau of Standards Miscellaneous Publication No. 271; Revision of its Circular No. 567.)

The general design and construction of instruments is treated in:

a. Elliott, A., and Dickson, J. H. *Laboratory Instruments; Their Design and Application.* (Second Edition.) New York: Chemical Publishing Company, 1960. 514 pp.

Some of the chapters are: The accuracy attainable in machining operations; Casting and jointing of metals; Constrained motion and constraints; Magnification of small displacements; Sensitivity and errors of instruments; Damping; Tests for straightness, flatness and squareness; Optical instruments.

b. Whitehead, T. N. *The Design and Use of Instruments and Accurate Mechanism.* New York: The Macmillan Company, 1934. 283 pp.

"This book is concerned with that type of mechanism whose function is directly dependent on the accuracy with which the component parts achieve their required relationships."

c. Scientific Instrument Manufacturers' Association, and British Scientific Instrument Research Association. *A Guide to Instrument Design.* London: Taylor and Francis, 1963. 444 pp.

Best design practice: mechanical, optical, electronic, etc.

Over thirty pieces of apparatus developed at various colleges can be duplicated economically through the *Apparatus Drawings Project* boxed sheets and book, published in 1962 by Plenum Press for the American Association of Physics Teachers and the American Institute of Physics.

First-hand acquaintance with current instrument output may be derived from annual exhibitions, either by personal attendance or through their descriptive materials. In the United States there is the annual physics show at the combined meeting of the American Physical Society and the American Association of Physics Teachers. The Instrument Society of America and the National Electronics Conferences also stage annual exhibits. In England there are annual exhibitions which put forth elaborate descriptive handbooks under the auspices of the Institute of Physics and the Physical Society (merged 1960).

Useful trade literature includes instrument catalogs, house organs, manuals, and more extensive publications like:

a. *Optical Industry and Systems Directory, 1967.* Pittsfield, Mass.: Optical Publishing Company, 1967. 500 pp.
Domestic and foreign manufacturers and suppliers, by name, products and location.

b. *British Instruments Directory and Data Handbook, 1965.* London: Scientific Instrument Manufacturers' Association, etc., 1965. Var. paged.

Relevant periodicals include the *Review of Scientific Instruments;* its British counterpart, the *Journal of Scientific Instruments; Journal of Applied Physics; Instruments and Control Systems; Electronics;* etc. (The last two publish annual directories.)

Laboratory Procedures

Effective methods of experimental attack and manipulation, that have been evolved over long periods through trial and error, are passed along in the form of procedure manuals. These aid the practicing scientist engaging in work outside his specialty, but are usually oriented towards the physics student in regular courses.

Experimental methodology is surveyed by the following multi-volume series:

Marton, Ladislaus L., editor-in-chief. *Methods of Experimental Physics.* New York: Academic Press, 1959- 7 vols., in process.
Volumes are titled: 1, Classical methods; 2, Electronic methods; 3, Molecular physics; 4 and 7, Atomic and electron physics; 5, Nuclear physics; and 6, Solid state physics.

Basic techniques.

Mastery of basic laboratory operations is essential before undertaking advanced experimental research. The following books are helpful:

a. Angerer, E. V., and Ebert H. *Physical Laboratory Handbook,*

translated from the German by W. Summer. Princeton, N. J.: D. Van Nostrand Company, 1966. 610 pp.
Materials, methods and techniques are covered in this comprehensive handbook. Some of the chapters are: Soldering and welding; Working of glass; Metal coatings; High-pressure techniques; Vacuum techniques; Electron microscopy; Optical procedures; Controls; Thermal instruments; Chronographs; Particle detectors; Lasers; Setting up the laboratory; and Safety in the laboratory.

b. Strong, John, *et al. Procedures in Experimental Physics.* New York: Prentice-Hall, Inc., 1938. 642 pp.
Wide range of data for the practicing physicist, e.g.: Fundamental operations in laboratory glass blowing; Laboratory optical work; Technique of high vacuum; Coating of surfaces; Electrometers and electroscopes; Geiger counters; Vacuum thermopiles and the measurement of radiant energy; Photoelectric cells and amplifiers; Molding and casting.

c. Kohlrausch, F. *Praktische Physik, zum Gebrauch für Unterricht, Forschung und Technik.* (21. Auflage.) Stuttgart: B. G. Teubner, 1960-1962. 2 vols.
This is an important collection of instrumental measurements and techniques, for research worker and student. It is the work of numerous editors, and has bibliographical references throughout.

d. Ansley, A. J. *An Introduction to Laboratory Technique.* (Second Edition.) London: The Macmillan Company, 1950. 288 pp.
The laboratory assistant in educational institutions will find this useful for suggestions on maintaining and constructing apparatus. It treats electrical measuring, electroplating, glass-working, graduation of apparatus, soldering, etc.

e. Institute of Physics and the Physical Society. *Laboratory and Workshop Notes,* compiled by Ruth Lang. London: Edward Arnold, Ltd., 1949- Vol. 1- (In progress.)
These are useful selections from the *Journal of Scientific Instruments.* Vol. 1 spanned 1923-1946; Vol. 7: 1962-1964.

For glass, a constantly used medium, there are:

a. Barr, W. E., and Anhorn, V. J. *Scientific and Industrial Glass Blowing and Laboratory Techniques.* Pittsburgh: Instruments Publishing Company, 1949. 388 pp.

b. Heldman, Julius D. *Techniques of Glass Manipulation in Scientific Research.* New York: Prentice-Hall, Inc., 1946. 132 pp.

c. Robertson, Andrew J. B., *et al. Laboratory Glass-Working for Scientists.* New York: Academic Press, 1957. 184 pp.

d. Dévé, Charles. *Optical Workshop Principles.* London: Adam Hilger, Ltd., 1943. 306 pp.

e. Twyman, Frank. *Prism and Lens Making.* (Second Edition.) London: Hilger and Watts, Ltd., 1952. 629 pp.

General manuals.

As will be indicated under Educational Approach (Chapter VII), modern educational trends are away from rote-learning towards purposeful action. Accordingly, students should be encouraged to think during the course of an experimental study, rather than follow the manual blindly as a routine task. Laboratory guides for college use are as numerous as the various textbooks of physics. The contents and arrangement of each reflect the author's preferences as to sequence, inclusion and style. Johnson's statement still holds:

> In the manuals now in use in school laboratories over the country the forms of presentation of subject matter reach two extremes. Some of these works—offered for standard college use—afford little more than child's play. Others—not infrequently the productions of men eminent for research—are so exacting, so mathematical, or so detailed, that they require uncommon preparation, and often a special equipment. They may be serviceable in the laboratories of the technical schools in which they were written, but they hardly seem suitable for general use.[7]

Only a few of the more adaptable general laboratory manuals need be cited:

a. Ingersoll, Leonard R.; Martin, Miles J.; and Rouse, T. A. *A Laboratory Manual of Experiments in Physics.* (Sixth Edition.) New York: McGraw-Hill Book Company, 1953. 286 pp.

b. White, Marsh W., and Manning, Kenneth V. *Experimental College Physics.* (Third Edition.) New York: McGraw-Hill Book Company, 1954. 347 pp.

The purposes of laboratory work were aptly summarized:

> (1) to fix more clearly in mind the great facts and principles of Nature being studied in the classroom, (2) to develop the student's thinking and reasoning powers, (3) to furnish an opportunity for a firsthand study and confirmation of some of the fundamental laws of the science, and (4) to acquaint the student with some of the methods, instruments and techniques of physical measurements, thus enabling him more adequately to obtain an appreciation of the possibilities and limitations of the scientific spirit and method of investigation.

---

7. E. H. Johnson, *Laboratory Physics,* p. iii. New York: Harcourt, Brace and Company, 1923.

c. Berkeley Physics Laboratory.[8] *Laboratory Physics.* New York: McGraw-Hill Book Company, 1964-1966. Parts A-D, in 3 vols.

d. Nelson, Edward B. *Physics Laboratory Manual.* (Second Edition.) Dubuque, Iowa: William C. Brown Company, 1964. 132 pp.

e. McCormick, W. Wallace. *Laboratory Experiments in Physics.* New York: The Macmillan Company, 1966. 192 pp.

f. Physical Science Study Committee. *Laboratory Guide for Physics.* Boston: D. C. Heath and Company, 1960. 87 pp.
For the new high school physics course.

General laboratory manuals of more advanced type are represented by the following group:

a. Brown, Thomas B., editor. *The Lloyd William Taylor Manual of Advanced Undergraduate Experiments in Physics.* Reading, Mass.: Addison-Wesley Publishing Company, 1959. 550 pp.
This memorial volume, prepared by the American Association of Physics Teachers, describes experiments and techniques for the guidance of students and teachers.

b. Mark, Hans, and Olson, N. T. *Experiments in Modern Physics.* New York: McGraw-Hill Book Company, 1966. 300 pp.
Some chapter headings: Measurement of atomic masses; Detection and measurement of ionizing radiations; Solid-state electronics; Nuclear magnetic resonance; and Lasers.

c. Melissinos, Adrian C. *Experiments in Modern Physics.* New York: Academic Press, 1966. 559 pp.
Among the chapters: Experiments in quantization; Quantum-mechanical systems; Radiation and particle detectors; Scattering experiments; Magnetic resonance experiments; and High-resolution spectroscopy.

d. Calthrop, J. E. *Advanced Experiments in Practical Physics.* (Second Edition.) London: William Heinemann, 1952. 142 pp.
Included are interesting advanced experiments of classical and modern physics, performed with standard equipment in most instances.

e. Worsnop, Bernard L., and Flint, Henry T. *Advanced Practical Physics for Students.* (Ninth Edition.) London: Methuen and Company, 1951. 754 pp.
This compendium is of unusually large proportions, and presents theoretical summaries and alternative procedures.

f. Jerrard, H. G., and McNeill, D. B. *Theoretical and Experi-*

8. See also A. M. Portis, "The Berkeley Physics Laboratory." *American Journal of Physics,* 32: 458-464, June 1964.

mental Physics. London: Chapman and Hall, Ltd., 1960. 624 pp. Sufficient theory has been included for understanding the experiments on elasticity, surface tension, heat transfer, acoustics, optics, electronics, etc.

See also the demonstration experiment compendia listed in the section Lectures under Presentational Approach in Chapter IX.

Special manuals.

Laboratory manuals and technique résumés for special fields of physics are helpful to those wishing to go beyond the general laboratory course. These may be supplemented by the general manuals, and by experiment sections found in certain books, e.g., W. C. Michels' *Electrical Measurements;* R. L. Weber's *Heat and Temperature Measurement;* F. R. Watson's *Sound*, and R. B. Hastings' *Physics of Sound*.

Aerodynamics

a. Pope, Alan Y., and Harper, John J. *Low-Speed Wind Tunnel Testing*. New York: John Wiley and Sons, 1966. 457 pp.

b. Pope, Alan Y., and Goin, Kennith L. *High-Speed Wind Tunnel Testing*. New York: John Wiley and Sons, 1965. 474 pp.

See also Fluid mechanics in index.

Atoms; Radiation

a. Brown, B. *Experimental Nucleonics*. Englewood Cliffs, N. J.: Prentice-Hall, Inc., 1963. 245 pp.
Discussion of techniques, with experiments on neutron detection, radioactivity, etc.

b. Valente, Frank A. *A Manual of Experiments in Reactor Physics*, prepared under the auspices of the Division of Technical Information, USAEC. New York: The Macmillan Company, 1963. 335 pp.
Background material and twenty-three experiments for colleges having nuclear reactors.

c. Glower, Donald D. *Experimental Reactor Analysis and Radiation Measurements*. New York: McGraw-Hill Book Company, 1965. 348 pp.
Graded experiments illustrate principles and techniques.

d. United States. Argonne National Laboratory. *Nuclear Reactor Experiments*, edited by J. Barton Hoag. Princeton, N. J.: D. Van Nostrand Company, 1958. 480 pp.

e. Harnwell, Gaylord P., and Livingood, J. J. *Experimental Atomic Physics*. New York: McGraw-Hill Book Company, 1933. 472 pp.

Modern atomic physics and measurement of its fundamental quantities.

f. Segrè, Emilio. *Experimental Nuclear Physics.* New York: John Wiley and Sons, 1953-1959. 3 vols.
Comprehensive survey of techniques.

g. Friedrich, A.; Langeheine, H.; and Ulbricht, H. *Experiments in Atomic Physics.* London: John Murray, 1966. 106 pp.
Like the *Science Masters' Books,* intended for teachers. Some chapters: Spectra; Radiation: laws of waves and particles; Determination of the constants h, e, c; Fluorescence and photoelectricity; Impact, spin, and interference of electrons; and Nuclear physics.

See also index under Radiation and Atomic Energy.

Cryophysics

a. Hoare, F. E.; Jackson, L. C.; and Kurti, N., editors. *Experimental Cryophysics.* London: Butterworths Scientific Publications, 1961. 388 pp.
Included are the technical problems encountered in setting up a laboratory for low-temperature production, measurement and research.

b. Mackinnon, Lachlan. *Experimental Physics at Low Temperatures.* Detroit, Mich.: Wayne State University Press, 1966. 273 pp.

c. White, Guy K. *Experimental Techniques in Low-Temperature Physics.* London: Oxford University Press, 1959. 328 pp.

d. Donnelly, R. J. *Experimental Superfluidity.* Chicago: University of Chicago Press, 1967. 264 pp.

See also Temperature (high and low) in the index.

Crystallography

Wooster, William A. *Experimental Crystal Physics.* London: Oxford University Press, 1957. 115 pp.

See also index under Crystallography.

Elasticity

Searle, George F. C. *Experimental Elasticity.* (Second Edition.) Cambridge, England: At the University Press, 1933. 189 pp.

See also Viscosity, and Properties (mechanical).

Electricity; Magnetism

a. Tinnell, Richard W. *Experiments in Electricity—Alternating Current.* New York: McGraw-Hill Book Company, 1966. 141 pp.

b. Tinnell, Richard W. *Experiments in Electricity—Direct Current.* New York: McGraw-Hill Book Company, 1966. 144 pp.

c. Gregg, Robert Q.; Hammond, H. E.; and Frost, R. H. *A Manual of Electrical Measurements.* Cambridge, Mass.: Addison-Wesley Press, 1950. 150 pp.
This is a book of experiments grouped under section headings such

as: Galvanometer; Resistance and capacitance measurements; Magnetic measurements; Vacuum tubes, etc.

d. Haus, Hermann A., and Penhune, John P. *Case Studies in Electromagnetism.* New York: John Wiley and Sons, 1960. 336 pp.
Problems with solutions are followed by laboratory experiments: Static fields; Skin effect; Frequency behavior of a resistor and a capacitor; Spherical coil; and Electromechanical energy converter.

e. Craggs, J. D., and Meek, J. M. *High Voltage Laboratory Technique.* London: Butterworths Scientific Publications, 1954. 404 pp.

See also index under Electricity.

Electronics

a. Müller, Ralph H.; Garman, R. L.; and Droz, M. E. *Experimental Electronics.* New York: Prentice-Hall, Inc., 1942. 330 pp.
In conjunction with characteristics and non-communication applications of electron tubes, seventy experiments are described in detail.

b. Schulz, E. H.; Anderson, L. T.; and Leger, R. M. *Experiments in Electronics and Communication Engineering.* (Second Edition.) New York: Harper and Brothers, 1954. 342 pp.
Covers circuits, electronics, ultra-high frequency techniques, etc.

c. Evans, Walter H. *Experiments in Electronics.* Englewood Cliffs, N. J.: Prentice-Hall, Inc., 1959. 374 pp.

Two compendia of techniques rather than experiments are:

a. Bachman, C. H. *Techniques in Experimental Electronics.* New York: John Wiley and Sons, 1948. 252 pp.
Equipment and procedures for maintaining beams of charged particles in a vacuum are discussed, under chapter headings such as: Pumps; Vacuum gauges; Controls and gadgets; Leak detection; Assembling and processing of electronic devices.

b. Elmore, William C., and Sands, Matthew. *Electronics; Experimental Techniques.* New York: McGraw-Hill Book Company, 1949. 417 pp.
The electronic instrumentation work that had been done at Los Alamos Laboratory during the development of the atomic bomb is described, under such headings as: Circuit components; Voltage amplifiers; Electronic counters; and Oscillographs.

See also index under Electronics.

Harmonic motion

Searle, George F. C. *Experimental Harmonic Motion.* Cambridge, England: At the University Press, 1922. 100 pp.
Physics teachers will agree with Searle when he states (p. 5):

> The subject of Harmonic Motion presents difficulties to many students. For some reason they fail to get any real grasp of the principles and in consequence dare not trust themselves to apply them to the simple examples they meet with in practical physics, even in those cases where the mathematical analysis is quite elementary. The present little volume is an attempt to meet the difficulty.

Optics

a. Palmer, Charles H. *Optics: Experiments and Demonstrations.* Baltimore: The Johns Hopkins Press, 1962. 321 pp.

b. Wagner, Albert F. *Experimental Optics.* New York: John Wiley and Sons, 1929. 203 pp.
Some of the divisions are: Index of refraction and spectrometry; Lens aberrations; Photometry; Telescopic instruments; Measurement of constants of microscopes.

c. Searle, George F. C. *Experimental Optics.* Cambridge, England: At the University Press, 1925. 357 pp.

d. Wood, Elizabeth A. *Experiments with Crystals and Light.* New York: Bell Telephone Laboratories, Inc., 1964. 109 pp.
Companion volume to her *Crystals and Light* for high schools; delightful optical effects produced.

See also index under Optics and Light.

Spectroscopy

a. Sawyer, Ralph A. *Experimental Spectroscopy.* (Third Edition.) New York: Dover Publications, 1963. 358 pp.
The text was modified and the bibliography updated. This is a survey of techniques rather than an experiment manual. It is designed to acquaint students and research workers with spectrographs and their use. There is an excellent descriptive list of spectroscopic charts and tables.

b. Harrison, George R.; Lord, R. C.; and Loofbourow, J. R. *Practical Spectroscopy.* New York: Prentice-Hall, Inc., 1948. 605 pp.

c. Tolansky, Samuel. *High Resolution Spectroscopy.* London: Methuen and Company, 1947. 291 pp.

See also index under Spectroscopy.

General measurements.

The importance of measurement in physical science is emphasized by Lord Kelvin:

> I often say that when you can measure what you are speaking about, and express it in numbers, you know something about it; but when you cannot measure it, when you cannot express it in numbers, your knowledge is a meagre and unsatisfactory kind: it may be the begin-

ning of knowledge, but you have scarcely, in your thoughts, advanced to the stage of *science,* whatever the matter may be.[9]

Measurement in general is discussed in:

a. Campbell, Norman R. *An Account of the Principles of Measurement and Calculation.* London: Longmans, Green and Company, 1928. 293 pp.

b. Baird, D. C. *Experimentation; An Introduction to Measurement Theory and Experiment Design.* Englewood Cliffs, N. J.: Prentice-Hall, Inc., 1962. 198 pp.
The physicist's basic philosophy of experimental measurement is developed.

c. Churchman, C. West, and Ratoosh, Philburn. *Measurement: Definitions and Theories.* New York: John Wiley and Sons, 1959. 274 pp.
Contents: Some meanings of measurement; Some theories of measurement; Some problems in the physical sciences; Some problems in the social sciences.

A few manuals on errors and data treatment follow:

a. Beers, Yardley. *Introduction to the Theory of Errors.* (Second Edition.) Reading, Mass.: Addison-Wesley Publishing Company, 1957. 66 pp.

b. Topping, J. *Errors of Observation and Their Treatment.* London: Institute of Physics, 1955. 119 pp.

c. Pugh, Emerson M., and Winslow, George H. *The Analysis of Physical Measurements.* Reading, Mass.: Addison-Wesley Publishing Company, 1966. 246 pp.

d. Jánossy, Lewis. *Theory and Practice of the Evaluation of Measurements.* Oxford: At the Clarendon Press, 1965. 481 pp.

e. Young, Hugh D. *Statistical Treatment of Experimental Data.* New York: McGraw-Hill Book Company, 1962. 172 pp.

f. Mandel, John. *Statistical Analysis of Experimental Data.* New York: Interscience Publishers, 1964. 410 pp.

Dimensional theory concerns the analysis of all physical quantities in terms of the fundamental entities, length, mass and time, irrespective of units employed. Helpful discussions include:

a. Lanchester, Frederick W. *The Theory of Dimensions and its Applications for Engineers.* London: Crosby Lockwood and Son, 1936. 314 pp.

---

9. W. T. Kelvin, *Popular Lectures and Addresses by Sir William Thomson,* Vol. 1, pp. 73-74. London: The Macmillan Company, 1889.

Written in most readable style, it has strong words on the controversial[10] "slug" and a good comparison of mass vs. weight.

b. Bridgman, Percy W. *Dimensional Analysis.* (Revised Edition.) New Haven, Conn.: Yale University Press, 1931. 113 pp.

c. Huntley, H. E. *Dimensional Analysis.* New York: Rinehart and Company, 1955. 158 pp.

d. Ipsen, D. C. *Units, Dimensions, and Dimensionless Numbers.* New York: McGraw-Hill Book Company, 1960. 236 pp.
Basic dimensions of length, mass, time, temperature and electric charge were chosen to cover mechanics, heat and electricity.

Selected papers from the U. S. National Bureau of Standards survey exact measuring:

Precision Measurement and Calibration, compiled by S. F. Booth. Washington, D. C.: Government Printing Office, 1961. 3 vols. (NBS Handbook No. 77.)
Volume titles are: 1, Electricity and electronics; 2, Heat and mechanics; and 3, Optics, metrology and radiation.

Another NBS publication charts the development of American standards for weights and measures[11] from colonial times to 1962: *Weights and Measures Standards of the United States; A Brief History,* compiled by Lewis V. Judson. Washington, D. C.: Government Printing Office, 1963. 30 pp. (NBS Miscellaneous Publication No. 247.)

The art of commercial weighing in early times is beautifully depicted in:

Kisch, Bruno. *Scales and Weights; A Historical Outline.* New Haven, Conn.: Yale University Press, 1965. 297 pp.

Weights and measures of two systems are compared in:

a. Kayan, Carl F., editor. *Systems of Units: National and International Aspects.* Washington, D. C.: American Association for the Advancement of Science, 1959. 297 pp. (Publication No. 57.) See

10. See also: L. A. Hawkins and S. A. Moss, "Alice and the Sluggers." *American Journal of Physics,* 13: 409-411, December 1945; and S. L. Gerhard, "Slugging Out a Case for the Pounders." *American Journal of Physics,* 18: 302-305, May 1950.

11. See also *Index to the Reports of the National Conference on Weights and Measures,* from the 1st to the 45th, 1905-1960. Washington, D. C.: Government Printing Office, 1962. 74 pp. (NBS Miscellaneous Publication No. 243.) For archeological and historical evidence concerning weights and measures, see A. E. Berriman, *Historical Metrology.* London: J. M. Dent and Sons, 1953. 224 pp.

particularly pp. 121-130, "Modern Physics Has Its Unit Problems," by W. W. Havens, Jr.

b. National Industrial Conference Board. *The Metric Versus the English System of Weights and Measures.* New York: The Century Company, 1921. 261 pp. (Its Research Report No. 42.)

See especially pp. 55-65 concerning use in science and engineering.

c. *Units of Weight and Measure (United States Customary and Metric); Definitions and Tables of Equivalents.* Washington, D. C.: Government Printing Office, 1946. 65 pp. (U. S. National Bureau of Standards Miscellaneous Publication No. 121.)

English units are favored in:

Ingalls, Walter R. *Units of Weights and Measures.* New York: American Institute of Weights and Measures, 1946. 49 pp.

The metric system is strongly advocated by:

National Council of Teachers of Mathematics. *The Metric System of Weights and Measures.* New York: Bureau of Publications, Teachers College, Columbia University, 1948. 303 pp. (Its *Twentieth Yearbook,* compiled by the Committee on the Metric System.)

Physicists have long made general use of metric units in their cgs system, based on fundamental units of the centimeter, gram, and second, and reviewed in:

Everett, Joseph D. *Illustrations of the C. G. S. System of Units.* (Fifth Edition.) London: The Macmillan Company, 1902. 283 pp.

More general in scope are:

a. Jerrard, H. G., and McNeill, D. B. *A Dictionary of Scientific Units.* (Second Edition.) London: Chapman and Hall, Ltd., 1964. 204 pp.

b. Clason, W. E. *Lexicon of International and National Units.* Amsterdam, New York, etc.: Elsevier Publishing Company, 1964. 75 pp.

Definitions and multilingual equivalents are given for international units; national units are listed with U. S., British, and metric values.

c. Hvistendahl, H. S. *Engineering Units and Physical Quantities.* London: The Macmillan Company, 1964. 132 pp.

d. Chertov, A. G. *Units of Measurement of Physical Quantities,* revised by Herbert J. Eagle. New York: Hayden Book Company, 1964. 165 pp.

For electricity and magnetism, the mks system was adopted by the International Electrotechnical Commission at Brussels in 1935, to take effect five years later. This application of meter-kilogram-second units (instead of the former arbitrary "international" units) is described in:

a. Jauncey, G. E. M., and Langsdorf, A. S. *M. K. S. Units and Dimensions.* New York: The Macmillan Company, 1940. 62 pp.

b. Sas, R. K., and Pidduck, F. B. *The Metric-Kilogram-Second System of Electrical Units.* London: Methuen and Company, 1947. 60 pp.

c. "What is the Meter-Kilogram-Second System of Units?" (A Report of the American Association of Physics Teachers Committee on Electric and Magnetic Units.) *American Physics Teacher,* 6: 144-151, June 1938.

d. Moon, Parry H., and Spencer, Domina E. "Utilizing the Mks System." *American Journal of Physics,* 16: 25-38, January 1948.

> The mks system gives a single comprehensive and international set of units for all of physics and engineering. . . . It has been employed extensively in recent textbooks, but authors seem to feel the lack of a general treatment of the mks system and a dearth of data expressed in mks units. The present paper attempts to remedy this difficulty by outlining the use of mks units in physics, and by giving tables of constants and conversion factors.

Standards may designate physical entities,[12] such as the pound-mass of a certain platinum block preserved in London, or a standard yard marked off on a bar, but also may refer to codified procedures, definitions, specifications, etc., adopted for purposes of uniformity and mutual understanding. For an interesting account of the struggle for scientific standards of measurement see:

Perry, John. *Story of Standards.* New York: Funk and Wagnalls Company, 1955. 271 pp.

For standards and their sponsors see:

a. Struglia, Erasmus J. *Standards and Specifications Information Sources: A Guide to Literature and to Public and Private Agencies Concerned with Technological Uniformities.* Detroit, Mich.: Gale Research Company, 1965. 187 pp.

b. Booth, Sherman F. *Standardization Activities in the United States; A Descriptive Directory.* Washington, D. C.: Government

12. For example, see these résumés of the work of our National Bureau of Standards: L. B. Macurdy, "Standards of Mass." *Physics Today,* 4 (No. 4): 7-11, April 1951; and R. E. Wilson, "Standards of Temperature." *Physics Today,* 6 (No. 1): 10-15, January 1953; also the Bureau's publications. British practice is reflected in: *Recent Developments and Techniques in the Maintenance of Standards.* London: H. M. Stationery Office, 1952. 100 pp.

Printing Office, 1960. 210 pp. (U. S. National Bureau of Standards Miscellaneous Publication No. 230.)

See also Houghton's *Technical Information Sources.*

Extensive series of standards representing American and English practice in many fields have been published by the United States of America Standards Institute (formerly American Standards Association), and by the British Standards Institution, with annual index lists. While mostly directed towards engineers, some USA titles impinge upon physics, such as:

Acoustical, colorimetric, electrical, electronic and photometric terminologies and measurement manuals; abbreviations of scientific and technical terms; and symbols for entities in acoustics, electricity, electronics, heat, mechanics, meteorology, photometry and physics. Each standard has an identifying number and year, e.g., S1.1-1960 (Acoustical Terminology).

A price list and index of USA Standards annually accompanies an issue of the USASI journal, *Magazine of Standards.* The institute issues descriptive material on the purposes and development of standards. It is active in the international field.

See also the multi-volume *Standards* and other publications of the American Society for Testing and Materials. (The word "and" was formerly not in the corporate name.)

Special measurements.

In addition to the books cited earlier in this chapter under Instruments, the following describe measurement techniques in various phases of physics experimentation:

Air flow

Ower, E. *The Measurement of Air Flow.* (Third Edition.) London: Chapman and Hall, Ltd., 1949. 293 pp.

This is based on work at the National Physical Laboratory.

See also Aerodynamics and Fluid Mechanics.

Color

a. *Handbook of Colorimetry,* prepared by the staff of the Color Measurement Laboratory, Massachusetts Institute of Technology, under the direction of Arthur C. Hardy. Cambridge, Mass.: The Technology Press, 1936. 87 pp.

The authors state (p. v.):

> Despite the age and extent of man's interest in color, colorimetry is a relatively new and unfamiliar science. Until physical instruments were developed which measure color in terms of quantities and wavelengths

of light, the only available methods of color specification had of necessity to be based on samples of the various colors. The fact that these samples are subject to change with time, even under the best of conditions, has made it impossible to accumulate an extensive and accurate body of knowledge concerning the diverse phenomena of color and color vision.

b. Wright, William D. *The Measurement of Colour.* (Third Edition.) Princeton, N. J.: D. Van Nostrand Company, 1964. 291 pp.

c. Bouma, Pieter J. *Physical Aspects of Colour.* New York: Elsevier Book Company, 1948. 312 pp.

d. *The ISCC—NBS Method of Designating Colors and a Dictionary of Color Names,* by Kenneth L. Kelly and Deane B. Judd. Washington, D. C.: Government Printing Office, 1955. 155 pp. (U. S. National Bureau of Standards Circular No. 553.) ISCC signifies Inter-Society Color Council.

The revised ISCC—NBS color designations are defined in Munsell terms by 31 name charts, one for each of 31 ranges of Munsell hue. The ISCC—NBS equivalents of 7500 color names previously defined by reference to 11 different sets of material standards have been determined and listed both alphabetically and by ISCC—NBS color designation to form a dictionary of color names.

e. *Colorimetry,* by Deane B. Judd. Washington, D. C.: Government Printing Office, 1950. 56 pp. (U. S. National Bureau of Standards Circular No. 478.)

See also index under Color.

Electricity; Electronics

a. Smith, Arthur W., and Wiedenbeck, M. L. *Electrical Measurements.* (Fifth Edition.) New York: McGraw-Hill Book Company, 1959. 307 pp.

b. Michels, Walter C. *Electrical Measurements and their Applications.* Princeton, N. J.: D. Van Nostrand Company, 1957. 331 pp. This supersedes his two earlier titles, and includes over forty experiments.

c. Harris, Forest K. *Electrical Measurements.* New York: John Wiley and Sons, 1952. 784 pp.

d. Stout, Melville B. *Basic Electrical Measurements.* (Second Edition.) Englewood Cliffs, N. J., Prentice-Hall, Inc., 1960. 571 pp.

e. Sucher, Max, and Fox, Jerome. *Handbook of Microwave Measurements.* (Third Edition.) New York: John Wiley and Sons, 1964. 3 vols.

f. Ginzton, Edward L. *Microwave Measurements.* New York: McGraw-Hill Book Company, 1957. 515 pp.

g. Terman, Frederick E., and Pettit, J. M. *Electronic Measurements.* (Second Edition.) New York: McGraw-Hill Book Company, 1952. 707 pp.

h. Prensky, Sol D. *Electronic Instrumentation.* Englewood Cliffs, N. J.: Prentice-Hall, Inc., 1963. 534 pp.

i. Thomas, Harry E., and Clarke, Carole A. *Handbook of Electronic Instruments and Measurement Techniques.* Englewood Cliffs, N. J.: Prentice-Hall, Inc., 1967. 416 pp.

See also index under Electricity and Electronics.

Light

a. Tolansky, Samuel. *An Introduction to Interferometry.* New York: Longmans, Green and Company, 1955. 223 pp.
"Interferometry is an elegant branch of optics. It is elegant because of its economy of means. By the correct use of very simple tools, it is possible to measure the size of a molecule and the diameter of a star."

b. Françon, Maurice. *Optical Interferometry.* New York: Academic Press, 1966. 307 pp.

c. Steel, W. H. *Interferometry.* Cambridge, England: At the University Press, 1967. 271 pp.

d. Tolansky, Samuel. *Surface Microtopography.* New York: Interscience Publishers, 1960. 296 pp.
With characteristic interest and enthusiasm, the author explains just how to use multiple-beam interferometry to measure surface irregularities.

e. Martin, A. E. *Infrared Instrumentation and Techniques.* New York: American Elsevier Publishing Company, 1966. 180 pp.
Infrared spectrometers and spectrophotometers are described.

f. Walsh, John W. T. *Photometry.* (Third Edition.) London: Constable and Company, 1958. 544 pp.
Includes bibliographical references 1925-1957, and selected earlier ones. For a complete bibliography of photometry to 1925 consult the first edition, 1926. Some of the chapters are: Radiation; The eye and vision; Photoelectric cells; Standards and sub-standards; Light, distribution and total flux measurement; Colorimetry; Spectrophotometry; The photometric laboratory.

Barrows' *Light, Photometry and Illuminating Engineering* is also useful.

See also index under Light, Color and Optics.

Pressure

Giardini, A. A., and Lloyd, Edward C., editors. *High-Pressure*

Measurement. London: Butterworths Scientific Publications, 1963. 417 pp.

Radiation; Radioactivity

a. Forsythe, W. E., editor. *Measurement of Radiant Energy.* New York: McGraw-Hill Book Company, 1937. 452 pp.
This National Research Council study "covers the radiation laws, the radiation constants, the care that should be taken with the source in radiation measurements, and the method of separating the radiation into wave-length intervals."

b. Smith, R. A.; Jones, F. E.; and Chasmar, R. P. *The Detection and Measurement of Infrared Radiation.* London: Oxford University Press, 1957. 458 pp.

c. Ritson, David M., editor. *Techniques of High Energy Physics.* New York: Interscience Publishers, 1961. 540 pp.
Detection and measurement of high energy particles.

d. Watt, D. E., and Ramsden, D. *High Sensitivity Counting Techniques.* Oxford, New York, etc.: Pergamon Press, 1964. 348 pp.
Alpha, beta and gamma radiation measurements.

e. Price, William J. *Nuclear Radiation Detection.* (Second Edition.) New York: McGraw-Hill Book Company, 1964. 430 pp.

f. Birks, J. B. *The Theory and Practice of Scintillation Counting.* Oxford, New York, etc.: Pergamon Press, 1964. 662 pp.

g. Dearnaley, Geoffrey, and Northrop, D. C. *Semiconductor Counters for Nuclear Radiations.* (Second Edition.) New York: John Wiley and Sons, 1966. 459 pp.

h. Snell, Arthur H., editor. *Nuclear Instruments and Their Uses.* New York: John Wiley and Sons, 1962-1965. 2 vols.
Nuclear detectors used in counting experiments, radiochemical work, and health physics.

See also index under Radiation and Radioactivity.

Sound

Beranek, Leo L. *Acoustic Measurements.* New York: John Wiley and Sons, 1949. 914 pp.
A useful section on terminology is included, pp. 15-36.

See also Chapter X—Sound.

Temperature; Heat

A monumental compendium on all aspects of temperature has been produced by the American Institute of Physics in cooperation with the U. S. National Bureau of Standards, the National Research Council, and many other organizations:

American Institute of Physics. *Temperature; Its Measurement and Control in Science and Industry.* New York: Reinhold Publishing Corporation, 1941-1963. 3 vols. in 5.
This indispensable set comprises papers presented at 1939, 1954 and 1961 symposia, supplemented by data tables, illustrations, etc. Theoretical concepts and measurement methods are fully covered. The third volume is in three parts.

Among shorter surveys are:

a. Weber, Robert L. *Heat and Temperature Measurement.* New York: Prentice-Hall, Inc., 1950. 422 pp.
This attractively designed book presents fundamental theory and a group of twenty-nine laboratory experiments, such as: Coefficient of expansion of a liquid; Radiation constant; Ratio of specific heats of air.

b. Hall, John A. *The Measurement of Temperature.* London: Chapman and Hall, Ltd., 1966. 96 pp.
This replaces earlier Institute of Physics monographs.

c. Griffiths, Ezer. *Methods of Measuring Temperature.* (Third Edition.) London: C. Griffin, 1947. 223 pp.

d. Swietoslawski, Wojciech. *Microcalorimetry.* New York: Reinhold Publishing Corporation, 1946. 199 pp.

For extremes of temperature, see Temperature (high and low) in the index.

Time

a. Whitrow, G. J. *The Natural Philosophy of Time.* Edinburgh: Thomas Nelson and Sons, 1961. 324 pp.
Of this McVittie states:

> There are few mathematical physicists who possess either the knowledge or the ability to investigate the nature of so basic a concept as that of time. Few physicists, indeed, are aware of the complexities inherent in the idea; and of the vast reading in the works of philosophers, psychologists, biologists, neurophysiologists, as well as in those of applied mathematicians, that an author must carry out before he can discuss time. However, this remarkable feat has been performed by G. J. Whitrow. . . . It may be confidently expected that the stimulus provided by this book will give rise to many new researches into the complex, and very basic, problem of time.[13]

b. Fraser, J. T., editor. *The Voices of Time: A Coöperative Survey of Man's Views of Time as Expressed by the Sciences and by the Humanities.* New York: George Braziller, Inc., 1966. 710 pp.

---

13. G. C. McVittie, in review. *American Journal of Physics,* 31: 312, April 1963.

c. Goudsmit, Samuel A., and Claiborne, Robert. *Time.* New York: Time, Inc., 1967. 200 pp. (Life Science Library.)
Some of the chapters of this colorful presentation: The elusive nature of time; Subdividing the year; Segments of the second; Einstein revolution; and Perfecting the clock.

d. Gold, T., and Schumacher, D. L. *The Nature of Time.* Ithaca, N. Y.: Cornell University Press, 1967. 260 pp.
Leading physicists met at Cornell to discuss the "phenomenon of time and to examine the possible ways of measuring and defining it."

For time measurement based on the cesium atomic standard see:

a. Hudson, George E. "Of Time and the Atom." *Physics Today,* 18 (No. 8): 34-38, August 1965.

b. Essen, Louis. "Accurate Measurement of Time." *Physics Today,* 13 (No. 7): 26-29, July 1960.

c. Richardson, John M., and Brockman, James F. "Atomic Standards of Frequency and Time." *The Physics Teacher,* 4: 247-256, September 1966.

Viscosity

a. Van Wazer, J. R., *et al.* *Viscosity and Flow Measurement; A Laboratory Handbook of Rheology.* New York: Interscience Publishers, 1963. 406 pp.

b. Merrington, A. C. *Viscometry.* London: Edward Arnold and Company, 1949. 142 pp.

See also Elasticity, and Properties (mechanical).

Experimental Research

Research programs conducted at research institutes, universities, industrial laboratories and elsewhere fulfill the dual purpose of expanding the frontiers of scientific knowledge, and of making possible the production of all the devices of contemporary civilization.

General.

Overall discussions of pure and applied research are provided by:

a. Freedman, Paul. *The Principles of Scientific Research.* (Second Edition.) Oxford: Pergamon Press, 1960. 227 pp.

b. Bush, George P., and Hattery, Lowell H. *Scientific Research: Its Administration and Organization.* Washington, D. C.: American University Press, 1950. 190 pp.

c. Cockcroft, Sir John, editor. *The Organization of Research*

Establishments. Cambridge, England: At the University Press, 1965. 275 pp.
Fifteen directors describe their complex research institutions.

d. Pelz, Donald C., and Andrews, Frank M. *Scientists in Organizations: Productive Climates for Research and Development.* New York: John Wiley and Sons, 1966. 318 pp.
Describes "kinds of working environments in which technical people were stimulated to high levels of creativity and performance."

e. *Work Activities and Attitudes of Scientists and Research Managers: Data from a National Survey.* Menlo Park, Calif.: Stanford Research Institute, 1965. 220 pp.

f. Suits, C. Guy. *Suits: Speaking of Research.* New York: John Wiley and Sons, 1965. 466 pp.
Essays and speeches on a wide variety of topics by an outstanding research scientist.

Research productivity is examined in:

a. Lipetz, Ben-Ami. *A Guide to Case Studies of Scientific Activity.* Carlisle, Mass.: Intermedia, Inc., 1965. 350 pp.
References to "authoritative, chronological accounts which cast light on the processes by which scientific and technical advances were made or were missed."

b. Lipetz, Ben-Ami. *The Measurement of Efficiency of Scientific Research.* Carlisle, Mass.: Intermedia, Inc., 1965. 262 pp.

c. Herwald, S. W. "Basic Research in Industry—Status Symbol or Necessity?" *Physics Today,* 15 (No. 1): 22-25, January 1962.

See also index under Industrial physics.

Incidentally, preoccupation with phases of organized research should not overshadow training of the embryo research student at college. Williams asserts:

> The experience of twenty years of undergraduate teaching has shown that college students may be successfully introduced to research work. If the rising generation is to play a leading part in the program of scientific research, much more attention must be paid in the future to arousing the interest of undergraduates in the various fields of science and so inspiring them with the spirit of research. The professions which college men and women follow when they leave college are to a large extent determined by the kind of work which has stimulated them in their undergraduate days.[14]

14. S. R. Williams, *Magnetic Phenomena,* p. ix. New York: McGraw-Hill Book Company, 1931.

Special.

Of great number and variety are the institutions specializing in different kinds of research. For descriptive data one may consult the manuals cited under Societies, and for historical:

Brauer, Ludolph, *et al. Forschungsinstitute; Ihre Geschichte, Organisation, und Ziele.* Hamburg: Paul Hartung, 1930. 2 vols. All countries and fields of research are covered. Some sections are in English, e.g., the descriptions of the Carnegie Institution of Washington, and the Smithsonian Institution.

A more recent directory-type "guide to establishments in Western Europe conducting, promoting or encouraging research in science and technology" has been published:

European Research Index. Guernsey, Channel Islands: Francis Hodgson Ltd., 1965. 2 vols.

Francis Hodgson Ltd. also plans a multi-volume guide to the structure of world science, geographically arranged, with descriptive and directory data.

American industrial research laboratories are listed in:

National Research Council. *Industrial Research Laboratories of the United States*, compiled by William W. Buchanan. (Twelfth Edition.) Washington, D. C.: Bowker Associates, 1965. 746 pp.

University and other non-profit research units may be found in:

Research Centers Directory, edited by Archie M. Palmer and Anthony T. Kruzas. (Second Edition.) Detroit: Gale Research Company, 1965. 666 pp.
Supplemented by quarterly *New Research Centers.*

Summaries of current research programs being conducted in American and British colleges, respectively, will be found in the annual *Directory of Engineering College Research and Graduate Study* (published as a supplement to the *Journal of Engineering Education* beginning March 1967); and *Scientific Research in British Universities and Colleges* (Department of Education and Science, and the British Council).

Individual research projects are detailed in a British journal, *Research Applied in Industry* (titled *Research* 1947-1957); in the three separate sections of the *Journal of Research* of our National Bureau of Standards, with summaries in its *Technical News Bulletin;* and in various journals such as the *Physical Review*, the *Journal of Applied Physics*, etc. Some laboratories, notably the National Physical Laboratory (London), have issued reprint and abstract series.

Summary

The scientific method requires that all hypotheses be tested by experimentation, most conveniently (but not exclusively) in the laboratory. Characteristic equipment and associated procedures are described in the books cited, which outline general techniques as well as those adapted to highly specialized areas. Measurement stands out as the fundamental criterion of exact science, and is applied to the various physical entities. The chapter closes with a glance at organized research, which nurtures new developments and puts them to work. Thus do pure and applied science merge.

CHAPTER VI

MATHEMATICAL APPROACH

Physical Constants and Mathematical Tools

Mathematics is not covered here except in its immediate relationship to physics, and other sources [1] must be consulted for more detailed treatment of that subject. Physical constants are numerical entities. They are tabulated in various compilations. These are followed by books that show how to apply mathematics to physics problems, beginning at the modest level of practice in elementary physics, and ascending to the imposing heights of advanced theoretical physics.

Overview

Physics, being an exact science, makes use of numbers and mathematical processes to indicate quantities and interrelationships. The level of difficulty and complexity ranges from simple manipulations of algebraic and trigonometrical formulae to the calculations associated with pure mathematics. Physicists are reluctant, however, to permit such mathematical tools to overshadow experimental techniques and meanings. Thus a well-known textbook is prefaced by the comment:

> The aim throughout the present work has been to treat the matters considered from a physical point of view, and particularly to avoid regarding the material as exercises in applied mathematics.[2]

Even an outstanding physicist like Faraday pleads:

> There is one thing I would be glad to ask you. When a mathematician engaged in investigating physical actions and results has arrived at his own conclusions, may they not be expressed in common language as fully, clearly and definitely as in mathematical formulae? If so, would it not be a great boon to such as we to express them so—translating them out of their hieroglyphics that we also might work upon them by experiment.[3]

---

1. Such as Parke's *Guide,* and Pemberton's; also G. A. Miller, *Historical Introduction to Mathematical Literature.* New York: The Macmillan Company, 1916. 302 pp.
2. F. C. Champion and N. Davy, *Properties of Matter,* p. v. London: Blackie and Son, 1936.
3. Letter from Faraday to Maxwell, quoted in A. C. Candler, *Atomic Spectra and the Vector Model,* p. v. (Second Edition.) Princeton, N. J.: D. Van Nostrand Company, 1964.

Complementary development of mathematics and science is traced in:

Bochner, Salomon. *The Role of Mathematics in the Rise of Science.* Princeton, N. J.: Princeton University Press, 1966. 386 pp.

The amount of mathematics utilized in undergraduate textbooks varies from practically none at all in certain very generalized "science surveys" to a maximum in advanced theoretical presentations. Some authors of introductory texts feel that calculus is indispensable, while others have expertly adapted the treatment to students lacking this tool.

Physical quantities are brought within the perspective of everyday life by:

a. Baravalle, Hermann von. *Zahlen für Jedermann aus Physik und Technik.* (New Edition.) Stuttgart: Franckh'sche Verlagshandlung, 1949. 158 pp.

Although usable as a compendium of numerical values of pure physics and technology, its popular appeal stems from interesting comparative grouping. Under Length we find two pages of values ranging from ultra-microscopic (0.000,004 mm.) to the extent of the Chinese Wall (2450 Km.); under Temperature are listed typical high values (e.g., electric oven 4000° C) down to the lowest obtainable (−273° C.). There is a comparative table of speed from walking to flying. Such tabulations stimulate visualization of physical quantities, especially when linked with class work by skillful teachers.

b. Houwink, R. *The Odd Book of Data.* New York: American Elsevier Publishing Company, 1965. 105 pp.

Partial contents are: The earth; Physics; Atoms and molecules; Energy; Oddly enough. Interesting comparisons are made, e.g.: one teaspoonful of water has at least as many molecules as the Atlantic contains teaspoonfuls of water.

Physical Constants

Tables of physical constants have been included under the Mathematical Approach because of their essentially numerical character. They often appear alongside strictly mathematical tables (functions, logarithms, etc.) in handbooks and compendia.

Handbooks.

Before the standard compendia receive our attention, it may well be centered upon "a new type of handbook, a kind of student's companion, which he could use to supplement his regular text":

Lindsay, Robert B. *Student's Handbook of Elementary Physics.* New York: Dryden Press, 1943. 382 pp.
Chapters on the various properties of matter are followed by: Illustrated dictionary of terms (pp. 135-270); Chronological table of physics (pp. 271-298); Bibliography of readings (pp. 299-308); Useful formulas (pp. 309-344); Physical constants and mathematical tables (pp. 345-369).

Comprehensive single-volume handbooks of physics are available:

a. *American Institute of Physics Handbook,* edited by Dwight E. Gray, *et al.* (Second Edition.) New York: McGraw-Hill Book Company, 1963. 2118 pp.
Over one-hundred contributors produced this reference collection of physical data, graphs, formulas, derivations, etc.

b. Condon, E. U., and Odishaw, Hugh, editors. *Handbook of Physics.* (Second Edition.) New York: McGraw-Hill Book Company, 1967. 1500 pp.
Theoretical exposition rather than merely tabulation of data characterizes this work.

Other compilations of physical and chemical data:

a. Chemical Rubber Company. *Handbook of Chemistry and Physics,* edited by Robert C. Weast. (Forty-Eighth Edition.) Cleveland, Ohio: The Company, 1967. Var. paged. (Revised annually.)

b. Lange, Norbert A., and Forker, Gordon M. *Handbook of Chemistry.* (Revised Tenth Edition.) New York: McGraw-Hill Book Company, 1967. 2001 pp.
Extensive changes have been incorporated in this revision of the 1961 edition.

c. Kaye, G. W. C., and Laby, T. H. *Tables of Physical and Chemical Constants and some Mathematical Functions,* edited by N. Feather, *et al.* (Thirteenth Edition.) New York: John Wiley and Sons, 1966. 249 pp.

d. *Smithsonian Physical Tables,* prepared by William E. Forsythe. (Ninth Edition.) Washington, D. C.: Smithsonian Institution, 1954. 827 pp.
Geophysical constants receive special emphasis.

e. Ardenne, Manfred, Baron von. *Tabellen zur Angewandten Physik: Elektronenphysik, Ionenphysik, Vakuumphysik, Kernphysik, Medizinische Elektronik, Hilfsgebiete.* (Zweite Auflage.) Berlin: Deutscher Verlag der Wissenschaften, 1962-1964. 2 vols.

Two handbooks which furnish fundamentals of physical science which engineers increasingly need:

a. Potter, James H., editor. *Handbook of the Engineering Sciences.* Vol. 1: *The Basic Sciences.* Princeton, N. J.: D. Van Nostrand Company, 1967. 1347 pp.
Main divisions are: Mathematics; Physics; Chemistry; Graphics; Statistics; Theory of experiments; and Mechanics. The second volume treats the applied sciences.

b. Eshbach, Ovid W., editor. *Handbook of Engineering Fundamentals.* (Second Edition.) New York: John Wiley and Sons, 1952. 1322 pp.
Some of the sections: Mathematical and physical tables; Physical units and standards; Mechanics; Electricity and magnetism; Radiation, light and acoustics; and Materials.

Handy reference compendia for students:

a. Robson, John. *Basic Tables in Physics.* New York: McGraw-Hill Book Company, 1967. 384 pp.

b. Childs, W. H. J. *Physical Constants Selected for Students.* (Eighth Edition.) London: Methuen and Company, 1934. 87 pp.

Basic formulas are found in:

a. Menzel, Donald H., editor. *Fundamental Formulas of Physics.* (Second Edition.) New York: Dover Publications, 1963. 2 vols.

b. Thomas, T. S. E. *Physical Formulae.* New York: John Wiley and Sons, 1953. 118 pp.

Evaluation of the so-called "constants" of physics is presented in:

a. Cohen, E. Richard; Crowe, Kenneth M.; and DuMond, Jesse W. M. *The Fundamental Constants of Physics.* New York: Interscience Publishers, 1957. 287 pp.

b. Sanders, J. H. *The Fundamental Atomic Constants.* (Second Edition.) New York: Oxford University Press, 1965. 98 pp.

c. Bearden, J. A., and Thomsen, John S. *A Survey of Atomic Constants.* Baltimore, Md.: The Johns Hopkins Press, 1955. 138 pp. *See also* their later résumé in *American Journal of Physics*, 27: 569-576, November 1959.

d. "New Values for the Physical Constants, As Recommended by the NAS-NRC." *Physics Today,* 17 (No. 2): 48-49, February 1964.

e. DuMond, Jesse W. M. "Pilgrims' Progress in Search of the Fundamental Constants." *Physics Today,* 18 (No. 10): 26-43, October 1965.

Comprehensive tables.

The Landolt-Börnstein series of physical-chemical tables constitutes the most extensive data compilation currently in progress. As the sixth edition is still in the process of being completed, reference is also made to the fifth edition, as well as to the "New Series" being published concurrently:

a. *Landolt-Börnstein Physikalisch-Chemische Tabellen.* (Fünfte Auflage.) Berlin: Springer, 1923-1936. 8 vols. as follows: Vols. 1-2, continuously paged, 1923; 1st Supp., 1927; 2nd Supp., Parts 1-2, continuously paged, 1931; 3rd Supp., Parts 1-3, continuously paged, 1935-1936.

In Vol. 1, pp. iv-xiv, may be found a table of contents for the entire set, and a full subject index appears in the last supplementary volume.

b. *Landolt-Börnstein Zahlenwerte und Funktionen aus Physik, Chemie, Astronomie, Geophysik und Technik,* herausgegeben von Arnold Eucken. (Sechste Auflage.) Berlin: Springer, 1950- 4 vols. in numerous parts, in process.

Volume titles are as follows: 1, Atom- und Molekularphysik; 2, Eigenschaften der Materie in ihren Aggregatzuständen; 3, Astronomie und Geophysik; and 4, Technik.

c. *Landolt-Börnstein Zahlenwerte und Funktionen aus Naturwissenschaften und Technik,* herausgegeben von K. H. Hellwege. (Neue Serie.) Berlin: Springer 1961- 6 "Groups" in numerous parts, in process.

The English titles of groups are: 1, Nuclear physics and technology; 2, Atomic and molecular physics; 3, Crystal and solid state physics; 4, Macroscopic and technical properties of matter; 5, Geophysics and space research; and 6, Astronomy, astrophysics and space research. The publisher explains the absence of a full-fledged seventh edition as follows:

> The "New Series," however, is not a revised or a supplementary new edition. The hitherto inflexible division has been discontinued. Instead of this, independent smaller volumes will be published in simple succession. Volumes dealing with closed fields will be supplemented only when necessary; fields which are being newly opened or evolving rapidly will be treated with greater frequency.

The earlier counterpart was:

National Research Council. *International Critical Tables of Numerical Data, Physics, Chemistry and Technology,* prepared under the auspices of the International Research Council and the National Academy of Sciences by the National Research Council of the United States of America. New York: McGraw-Hill Book Company, 1926-1933. 7 vols. plus index vol.

This selective tabulation of properties was compiled by many scientists in various countries. It is arranged according to a rather complicated system explained on page 96 of Volume 1, or more succinctly in Soule's guide.[4] One often finds formulas instead of direct values.

A series of annual tables formed the basis of the *International Critical Tables* and also supplemented them, as follows:

a. *Tables Annuelles de Constantes et Données Numériques de Chimie, de Physique et de Technologie,* années 1910-1934. Paris: Gauthier-Villars, 1912-1937. Vols. 1-11, Part 1; plus collective indexes to vols. 1-5 and 6-10.

Bibliographical mention [5] has been made of "Vol. 12, 1935-6; (One or two parts cover 1931-9)." There is also a related series of monographs.

b. *Annual Tables of Physical Constants and Numerical Data.* Princeton, N. J.: Frick Chemical Laboratory, 1941-1942. 2 vols.

This was compiled by the National Research Council's American Committee on Annual Tables. Photo-reproduced pages were accompanied by a card service. The project is inactive.

The National Research Council's *Continuing Numerical Data Projects,* 2nd ed. (Publication 1463, 1966), and its *Consolidated Index of Selected Property Values* (Publication 976, 1962) represent updating efforts.

A reference set stemming from the International Union of Pure and Applied Chemistry Congress held in London in 1947 appears under French and English titles, with bilingual introductions and notes, as follows:

Tables de Constantes et Données Numériques (*Constantes Sélectionées*). Paris: Hermann et Cie, 1947- ;

Tables of Constants and Numerical Data. Oxford, New York, etc.: Pergamon Press, 1947- (In progress.) Extensive bibliographies are appended to these volumes, which cover special subjects and vary in size from 41 pp. (Vol. 8, Oxidation-reduction potentials), to 1000 pp. (Vol. 14, Optical rotatory power; a revision of Vol. 6, twice as large.)

For solids with melting points above 1000° F there is:

Touloukian, Y. S., editor. *Thermophysical Properties of High Temperature Solid Materials.* New York: The Macmillan Company, 1967. 6 vols. in 9.

4. B. A. Soule, *Library Guide for the Chemist,* pp. 200-202. New York: McGraw-Hill Book Company, 1938.

5. "List of Compendia and Data Tables in Physics," prepared by the Royal Society. *Journal of Documentation,* 7: 252-255, December 1951.

(This is a revision of the five-volume *Handbook . . .* edited by A. Goldsmith, *et al.*, Pergamon-Macmillan, 1961-1963.)

Comprehensive coverage of thermophysical properties data is furnished by:

Touloukian, Y. S., editor. *Thermophysical Properties Research Literature Retrieval Guide.* (Second Edition.) New York: Plenum Press, 1967. 3 vols.
Included are: Thermal conductivity; Specific heat; Viscosity; Thermal radiative properties; Diffusion coefficient; Permeability; Thermal diffusivity; and Prandtl number.

See also index under Tables, nuclear; and H. T. Johnson and D. L. Grigsby's *Electronic Properties of Materials.*

Special.

The following are examples of specialized compilations:

Gases and vapors

a. Geyer, E. W., and Bruges, E. A. *Tables of Properties of Gases, with Dissociation Theory and its Applications.* New York: Longmans, Green and Company, 1948. 102 pp.

b. Keenan, Joseph H., and Kaye, Joseph. *Gas Tables; Thermodynamic Properties of Air, Products of Combustion, and Component Gases.* New York: John Wiley and Sons, 1948. 238 pp.

c. Keenan, Joseph H., and Keyes, Frederick G. *Thermodynamic Properties of Steam, Including Data for the Liquid and Solid Phases.* New York: John Wiley and Sons, 1936. 89 pp.

Metals

a. Smithells, Colin J., editor. *Metals Reference Book.* (Fourth Edition.) London: Butterworths Scientific Publications, 1967. 3 vols.
Critically selected data relating to metal physics and metallurgy.

b. *Metals Handbook,* edited by Taylor Lyman, *et al.* (Eighth Edition.) Novelty, Ohio: American Society for Metals, 1961- 5 vols., in process.

Mathematical Tables

As this is a guide primarily to physics literature, no attempt will be made to list mathematical tables in general, especially since this has been accomplished in:

a. Fletcher, Alan, *et al. An Index of Mathematical Tables.* (Second Edition.) Oxford: Blackwell, for Scientific Computing Service; Reading, Mass.: Addison-Wesley Publishing Company, 1962. 2 vols.
References in Part 1, which is arranged according to function, are to

the bibliography in Part II, listing available tables. This index is a great time saver, and points out mathematical compilations that otherwise might be overlooked. For example, one learns where to find π correct to over 16,000 decimals!

b. Lebedev, A. V., and Fedorova, R. M. *A Guide to Mathematical Tables,* prepared from the Russian by D. G. Fry; *Supplement 1,* by N. M. Burunova. Oxford, New York, etc.: Pergamon Press, 1960. 586 pp.; 190 pp.

This complements Fletcher, especially for Russian tables.

c. Schütte, Karl. *Index Mathematischer Tafelwerke und Tabellen.* München: R. Oldenbourg, 1955. 143 pp.

About 1200 tables in various fields of science and technology are listed, with English and German captions.

See also a convenient list in *American Institute of Physics Handbook,* 2nd ed., pp. I-2 to I-14.

Current information and review articles on mathematical tables appear in the National Research Council's quarterly *Mathematics of Computation,* formerly *Mathematical Tables and Other Aids to Computation.*

Computation aids.

Computers qualify as "computation aids" in the grand manner. Interesting comments, pro and con, on their acceptance by scientists may be found in:

Spinrad, Robert J. "The Computer and You." *Physics Today,* 18 (No. 12): 47-54, December 1965.

Among introductory sources of information are:

a. Alt, Franz L. *Electronic Digital Computers; Their Use in Science and Engineering.* New York: Academic Press, 1958. 336 pp.

b. McCormick, E. M. *Digital Computer Primer.* New York: McGraw-Hill Book Company, 1959. 214 pp.

c. Von Handel, Paul, editor. *Electronic Computers: Fundamentals, Systems, and Applications.* Englewood Cliffs, N. J.: Prentice-Hall, Inc., 1961. 235 pp.

d. Klerer, Melvin, and Korn, Granino A., editors. *Digital Computer User's Handbook.* New York: McGraw-Hill Book Company, 1967. Var. paged.

e. Desmonde, William H. *Computers and Their Uses.* Englewood Cliffs, N. J.: Prentice-Hall, Inc., 1964. 296 pp.

f. Hersee, E. H. W. *A Simple Approach to Electronic Computers.* (Second Edition.) New York: Gordon and Breach, 1967. 261 pp.

That computers may be utilized in teaching on a relatively small scale is shown by:

a. Ahearne, John F. "Introductory Physics Experiments Using a Digital Computer." *American Journal of Physics,* 34: 309-313, April 1966.

b. Hartman, Roger D. "Use of Computers in an Undergraduate Light and Optics Laboratory." *American Journal of Physics,* 34: 793-798, September 1966.

c. Shirer, Donald L. "Computers and Physics Teaching. Part 1: Digital Computers." *American Journal of Physics,* 33: 575-583, July 1965.

The Commission on College Physics has issued a handbook, *The Computer in Physics Instruction,* based on a 1965 University of California conference.

See also separate sub-sections on analog and digital computers in physics research, in *American Institute of Physics Handbook,* 2nd ed., pp. I-15 to I-56.

Tabular compilations designed to facilitate mathematical processes vary widely. For reducing drudgery in arithmetic there are:

a. Barlow, Peter. . . . *Tables of Squares, Cubes, Square Roots, Cube Roots, and Reciprocals of Numbers up to 12,500,* edited by L. J. Comrie. (Fourth Edition.) London: E. and F. N. Spon, 1941. 258 pp.

b. Allen, Edward S. *Six-place Tables.* (Seventh Edition.) New York: McGraw-Hill Book Company, 1947. 232 pp.
This is a pocket-sized compendium of squares, cubes, square roots, cube roots, logarithms, trigonometric functions, etc.

Conversion factors are helpful, as found in:

Zimmerman, O. T., and Lavine, Irvin. *Conversion Factors and Tables.* (Third Edition.) Dover, N. H.: Industrial Research Service, 1961. 680 pp.

For logarithms, the handiest sources of seven-place values are:

a. Vega, Georg. *Logarithmic Tables of Numbers and Trigonometrical Functions,* translated from the 40th revised edition. New York: D. Van Nostrand Company, 19—. 575 pp.

b. Bruhns, Karl. *A New Manual of Logarithms to Seven Places of Decimals.* (Eighteenth Stereotype Edition.) New York: D. Van Nostrand Company, 1939. 610 pp.
This also furnishes logarithms of trigonometric functions. Methods of working with logarithmic tables are described in the introduction, pp. xi-xxiii.

Trigonometric relations are available to four and seven places, respectively, in:

a. Oglesby, Ernest J., and Cooley, Hollis R. *Logarithmic and Trigonometric Tables.* New York: Prentice-Hall, Inc. 1930. 76 pp.

b. Ives, Howard C. *Natural Trigonometric Functions to Seven Decimal Places for Every Ten Seconds of Arc.* (Second Edition.) New York: John Wiley and Sons, 1945. 368 pp.

Subdivision is sufficiently minute for most purposes.

Various general mathematical compendia are available, such as:

a. Burington, Richard S. *Handbook of Mathematical Tables and Formulas.* (Fourth Edition.) New York: McGraw-Hill Book Company, 1965. 423 pp.

b. Clements, Guy R., and Wilson, Levi T. *Manual of Mathematics and Mechanics.* (Second Edition.) New York: McGraw-Hill Book Company, 1947. 349 pp.

c. *Handbook of Mathematical Tables,* edited by Robert C. Weast, *et al.* (Second Edition.) Cleveland, Ohio: Chemical Rubber Company, 1964. 680 pp.

This supplements their *Handbook of Chemistry and Physics.*

Functions.

Integrals are conveniently tabulated in:

a. Peirce, B. O. *A Short Table of Integrals.* (Fourth Edition.) Boston: Ginn and Company, 1956. 189 pp.

b. Dwight, Herbert B. *Tables of Integrals and Other Mathematical Data.* (Fourth Edition.) New York: The Macmillan Company, 1961. 336 pp.

c. Gradshteyn, I. S., and Ryzhik, I. M. *Table of Integrals, Series, and Products.* (Fourth Edition.) New York: Academic Press, 1965. 1086 pp.

The most frequently helpful general collections of mathematical functions are:

a. Jahnke, Eugene, and Emde, Fritz. *Tables of Higher Functions,* revised by Friedrich Lösch. (Sixth Edition.) Stuttgart: B. G. Teubner; New York: McGraw-Hill Book Company, 1960. 318 pp.

b. *Handbook of Mathematical Functions, with Formulas, Graphs and Mathematical Tables,* edited by Milton Abramowitz and Irene A. Stegun. Washington, D. C.: Government Printing Office, 1964. 1045 pp. (U. S. National Bureau of Standards, Applied Mathematics Series No. 55.)

This has been characterized as a modernized and expanded version of the preceding item. Dover Publications, N. Y., offers a cheaper soft-cover reprint (1965).

c. Dwight, Herbert B. *Mathematical Tables of Elementary and Some Higher Mathematical Functions.* (Second Edition.) New York: Dover Publications, 1958. 217 pp.

Special collections may be represented by:

a. Harvard University Computation Laboratory. *Tables of the Bessel Functions of the First Kind of Orders.* Cambridge, Mass.: Harvard University Press, 1947-1951. (Its *Annals,* Vols. 3-14.)

b. Gray, Andrew, and Mathews, G. B. *A Treatise on Bessel Functions*[6] *and Their Applications to Physics.* (Second Edition.) London: The Macmillan Company, 1922. 327 pp.

c. Roberts, G. E., and Kaufman, H. *Table of Laplace Transforms.* Philadelphia: W. B. Saunders, 1966. 367 pp.

Applied Mathematics

This section deals with physics applications of mathematics, over the extremely wide range from elementary problem solving to advanced theoretical physics.

Dictionaries often provide starting points:

a. *International Dictionary of Applied Mathematics,* edited by W. F. Freiberger. Princeton, N. J.: D. Van Nostrand Company, 1960. 1173 pp.

b. James, Glenn, and James, Robert C. *Mathematics Dictionary.* (Second Edition.) Princeton, N. J.: D. Van Nostrand Company, 1959. 546 pp.
Interrelated mathematical concepts rather than mere definitions are presented. Multilingual indexes are appended.

Compendia are also helpful:

a. Korn, Granino A., and Korn, Theresa M. *Mathematical Handbook for Scientists and Engineers; Definitions, Theorems, and Formulas for Reference and Review.* New York: McGraw-Hill Book Company, 1961. 943 pp.

b. Meyler, Dorothy S., and Sutton, O. G. *A Compendium of Mathematics and Physics.* Princeton, N. J.: D. Van Nostrand Company, 1958. 384 pp.

---

6. For a shorter survey, see T. A. Benham, "Bessel Functions in Physics." *American Journal of Physics,* 15: 285-294; 488-497, July-August and November-December, 1947.

Vector analysis is an important tool:

a. Moon, Parry H., and Spencer, Domina E. *Vectors.* Princeton, N. J.: D. Van Nostrand Company, 1965. 334 pp.

b. Barnett, Raymond A., and Fujii, John N. *Vectors.* New York: John Wiley and Sons, 1963. 132 pp.

c. Hoffmann, Banesh. *About Vectors.* Englewood Cliffs, N. J.: Prentice-Hall, Inc., 1966. 134 pp.

Problems.

Mathematical solution in general is discussed from several viewpoints by:

a. Dadourian, H. M. *How to Study; How to Solve (Arithmetic through Calculus).* (New Edition.) Cambridge, Mass.: Addison-Wesley Press, 1951. 121 pp.

b. Kogan, Zuce. *Essentials in Problem Solving.* Chicago: The Author, 1951. 79 pp.

c. Polya, George. *Mathematical Discovery: On Understanding, Learning, and Teaching Problem Solving.* New York: John Wiley and Sons, 1962-1965. 2 vols.

Physics problem books serve to relate theoretical principles to practical solutions with correct units, facilitating present learning and subsequent review.

For general physics, elementary and advanced, there are:

a. Sackheim, George I. *Physics Calculations.* New York: The Macmillan Company, 1960. 267 pp.

b. MacDonald, Simon G. G. *Problems and Solutions in General Physics; For Science and Engineering Students.* Reading, Mass.: Addison-Wesley Publishing Company, 1967. 276 pp.

c. Pippard, A. B. *Cavendish Problems in Classical Physics.* Cambridge, England: At the University Press, 1962. 51 pp.

d. Strelkov, S. P., *et al.* *Problems in Undergraduate Physics,* edited by Dirkter Haar. Oxford, New York, etc.: Pergamon Press, 1965. 4 vols.

Volume titles are: 1, Mechanics; 2, Electricity and magnetism; 3, Optics; and 4, Molecular physics, thermodynamics, atomic and nuclear physics.

e. Shaskol'skaya, M. P., and El'tsin, I. A. *Selected Problems in Physics with Answers.* Oxford, New York, etc.: Pergamon Press, 1963. 246 pp.

f. Cronin, Jeremiah A., *et al.* *University of Chicago Graduate*

Problems in Physics, With Solutions. Reading, Mass.: Addison-Wesley Publishing Company, 1967. 263 pp.

g. Lebedev, N. N.; Skal'sksya, I. P.; and Uflyand, Ya. S. *A Collection of Problems in Mathematical Physics.* Oxford, New York, etc.: Pergamon Press, 1966. 406 pp.

h. Misyurkeyev, I. V. *Problems in Mathematical Physics.* New York: McGraw-Hill Book Company, 1966. 160 pp.

i. Choquet-Bruhat, Y. *Problems and Solutions in Mathematical Physics.* San Francisco: Holden-Day, Inc., 1967. 314 pp.

A helpful overview of physics principles for developing facility in problem solving is:

Weber, Robert L. *Review of College Physics.* Englewood Cliffs, N. J.: Prentice-Hall, Inc., 1958. 3 vols.

Also useful are compilations of principles and problems in Daniel Schaum's outline series (Schaum Publishing Company) and in the College Outline Series (Barnes and Noble, Inc.).

Special problem books are available for certain fields, such as:

a. Meshcherskii, I. V. *A Collection of Problems of Mechanics,* translated from the 26th Russian edition. Oxford, New York, etc.: Pergamon Press, 1965. 518 pp.

b. Fried, R. *Introductory Physics: Problems and Solutions in Mechanics.* Boston: Allyn and Bacon, 1966. 231 pp.

c. Pincherle, L. *Worked Problems in Heat, Thermodynamics and Kinetic Theory for Physics Students.* London: Oxford University Press, 1966. 150 pp.

d. Irodov, I. Ye. *A Collection of Problems in Atomic and Nuclear Physics.* Oxford, New York, etc.: Pergamon Press, 1966. 239 pp.

e. Gol'dman, I. I., *et al. Selected Problems in Quantum Mechanics,* edited by D. ter Haar. (Second Edition.) New York: Academic Press, 1965. 402 pp.

f. Batygin, V. V., and Toptygin, I. N. *Problems in Electrodynamics.* New York: Academic Press, 1965. 504 pp.

g. Benson, Frank A. *Problems in Electronics with Solutions.* (Fourth Edition.) London: E. and F. N. Spon, 1965. 307 pp.

See also the worked problems in Haus and Penhune's *Case Studies in Electromagnetism.*

Theoretical physics.

When advanced mathematical reasoning is applied to physical theory so as to derive new concepts that can be verified experimentally, the process is termed mathematical or theoretical physics. Although its wonders may be obscured to many by the complicated mathematics involved, it is not too distant from reality, as Slater and Frank state:

> The same ability to overcome obstacles, the same ingenuity in devising one method of procedure when another fails, the same physical intuition leading one to perceive the answer to a problem through a mass of intervening detail, the same critical judgment leading one to distinguish right from wrong procedures, and to appraise results carefully on the grounds of physical plausibility, are required in theoretical and in experimental physics. Leaks in vacuum circuits or in electric circuits have their counterparts in the many disastrous things that can happen to equations. And it is often as hard to devise a mathematical system to deal with a difficult problem, without unjustifiable approximations and impossible complications, as it is to design apparatus for measuring a difficult quantity or detecting a new effect. These things cannot be taught. They come only from that combination of inherent insight and faithful practice which is necessary to the successful physicist. But half the battle is over if the student approaches theoretical physics not as a set of mysterious formulas, or as a dull routine to be learned, but as a collection of methods, of tools, of apparatus, subject to the same sort of rules as other physical apparatus, and yielding results of great importance.[7]

The development of mathematical or theoretical physics is ably traced by:

Abro, A. d'. *Rise of the New Physics; Its Mathematical and Physical Theories.* (Second Edition.) New York: Dover Publications, 1951. 2 vols.

This was formerly titled *Decline of Mechanism (in Modern Physics)*.

An interesting discussion of the role of the mathematical physicist is afforded by:

Milne, E. A. *The Aims of Mathematical Physics.* Oxford: At the Clarendon Press, 1929. 28 pp.

Excellent general surveys include:

a. Joos, Georg. *Theoretical Physics.*[8] (Third Edition.) New York: Hafner Publishing Company, 1958. 885 pp.

b. Lindsay, Robert B. *Concepts and Methods of Theoretical Physics.* New York: D. Van Nostrand Company, 1951. 515 pp.

c. Page, Leigh. *Introduction to Theoretical Physics.* (Third Edition.) New York: D. Van Nostrand Company, 1952. 701 pp.

---

7. J. C. Slater and N. H. Frank, *Introduction to Theoretical Physics*, p. vii. New York: McGraw-Hill Book Company, 1933.
8. The German edition is titled *Lehrbuch der Theoretischen Physik.* (Zehnte Auflage.) Frankfurt am Main: Akademische Verlagsgesellschaft, 1959. 842 pp.

d. Blass, Gerhard A. *Theoretical Physics.* New York: Appleton-Century-Crofts, Inc., 1962. 451 pp.

For more extensive multi-volume treatment, see:

a. Landau, L. D., and Lifshitz, E. M. *Course of Theoretical Physics,* translated from the Russian. Reading, Mass.: Addison-Wesley Publishing Company, 1958-1965. Vols. 1-3; 5-8, as follows: 1, Mechanics; 2, Classical theory of fields (Second Edition); 3, Quantum mechanics—non-relativistic theory (Second Edition); 5, Statistical physics; 6, Fluid mechanics; 7, Theory of elasticity; and 8, Electrodynamics of continuous media.
The fourth volume will round out quantum theory. G. E. Uhlenbeck characterized the *Course* as being in the great tradition, and the only one attempted by physicists of the present generation. (*American Journal of Physics,* 27: 372, May 1959.)

b. Sommerfeld, Arnold. *Lectures on Theoretical Physics,* translated from the German. New York: Academic Press, 1949-1956. 6 vols., as follows: 1, Mechanics; 2, Mechanics of deformable bodies; 3, Electrodynamics; 4, Optics; 5, Thermodynamics and statistical mechanics; and 6, Partial differential equations in physics.

Various Slater books constitute a similar series on theoretical physics, stemming from the 1933 *Introduction.*

Further help with applied mathematical techniques is obtainable from:

a. Margenau, Henry, and Murphy, George M. *The Mathematics of Physics and Chemistry.* (Second Edition of Vol. 1.) Princeton, N. J.: D. Van Nostrand Company, 1956-1964. 2 vols.
Coverage of formulas and methods in the first volume is extended by the twelve supplementary chapters of the second.

b. Courant, R., and Hilbert, D. *Methods of Mathematical Physics.* New York: Interscience Publishers, 1953-1962. 2 vols.
A third volume is in press. The second volume of this classic presentation treats partial differential equations.

c. Morse, Philip M., and Feshbach, Herman. *Methods of Theoretical Physics.* New York: McGraw-Hill Book Company, 1953. 2 vols.

d. Sokolnikoff, Ivan S., and Redheffer, R. M. *Mathematics of Physics and Modern Engineering.* (Second Edition.) New York: McGraw-Hill Book Company, 1966. 752 pp.

e. Jeffreys, Sir Harold, and Jeffreys, Bertha S. *Methods of Mathematical Physics.* (Third Edition.) Cambridge, England: At the University Press, 1956. 714 pp.

f. Polozhiy, G. N. *Equations of Mathematical Physics.* New York: Hayden Book Company, 1967. 543 pp.

Comprehensive collections of formulae useful in theoretical physics are presented in:

a. Madelung, Erwin R. *Die Mathematischen Hilfsmittel des physikers.* (Siebente Auflage.) Berlin: Springer, 1964. 535 pp.

b. Magnus, Wilhelm; Oberhettinger, F.; and Soni, R. P. *Formulas and Theorems for the Special Functions of Mathematical Physics.* (Third Edition.) New York: Springer, 1966. 508 pp.

Examples of special tools in mathematical physics are:

a. Berg, Ernst J. *Heaviside's Operational Calculus as Applied in Engineering and Physics.* (Second Edition.) New York: McGraw-Hill Book Company, 1936. 258 pp.

b. Churchill, Ruel V. *Fourier Series and Boundary Value Problems.* (Second Edition.) New York: McGraw-Hill Book Company, 1963. 248 pp.

c. Bateman, H. *Partial Differential Equations of Mathematical Physics.* Cambridge, England: At the University Press, 1932. 522 pp.

Statistical techniques are applied to physics in:

a. Lindsay, Robert B. *Introduction to Physical Statistics.* New York: John Wiley and Sons, 1941. 306 pp.

Lindsay states (p. v):

> In this book the attempt has been made to survey as thoroughly as possible the various ways in which statistical reasoning has been used in physics from the classical applications to fluctuation phenomena, kinetic theory, and statistical mechanics to the contemporary quantum mechanical statistics. Emphasis has been laid on methodology.

b. Wannier, Gregory H. *Statistical Physics.* New York: John Wiley and Sons, 1966. 532 pp.

c. Kittel, Charles. *Elementary Statistical Physics.* New York: John Wiley and Sons, 1958. 228 pp.

d. Haar, Dirk ter. *Elements of Thermostatistics.* (Second Edition.) New York: Holt, Rinehart and Winston, 1966. 316 pp.

See also Landau and Lifshitz, Vol. 5 (1958), which is a revision of their *Statistical Physics* (1938).

Further aid with statistical mathematics may be derived from Bacon's résumé,[9] and from such books as:

---

9. R. H. Bacon, "Practical Statistics for Practical Physicists." *American Journal of Physics,* 14: 84-98; 198-209, March-April and May-June, 1946.

a. Fisher, Ronald A. *The Design of Experiments.* (Fifth Edition.) New York: Hafner Publishing Company, 1949. 242 pp.

b. Aitken, A. C. *Statistical Mathematics.* (Fifth Edition.) New York: Interscience Publishers, 1947. 161 pp.

c. Burington, Richard S., and May, D. C., Jr. *Handbook of Probability and Statistics, with Tables.* Sandusky, Ohio: Handbook Publishers, Inc., 1953. 332 pp.

Statistical treatment of information "bits" flow-in communication channels has been largely taken over from the mathematicians by electrical engineers, biologists, librarians, *et al.*:

a. Raisbeck, Gordon. *Information Theory; An Introduction for Scientists and Engineers.* Cambridge, Mass.: The M.I.T. Press, 1964. 105 pp.

b. Brillouin, Léon. *Science and Information Theory.* (Second Edition.) New York: Academic Press, 1962. 351 pp.

c. Abramson, Norman. *Information Theory and Coding.* New York: McGraw-Hill Book Company, 1963. 201 pp.

Advanced mathematical treatments of physical phenomena are scattered among sections of this guide in the Topical Approach. However, two impressive areas of theoretical physics may herewith be mentioned as examples:

Quantum theory

Electron radiation possesses both wave and discrete-particle attributes, as outlined in:

a. Heitler, W. *Quantum Theory of Radiation.* (Third Edition.) London: Oxford University Press, 1954. 430 pp.

b. Frenkel, J. *Wave Mechanics; Elementary Theory.* (Second Edition.) Oxford: At the Clarendon Press, 1936. 312 pp. (This volume, and one on advanced theory, have been reprinted by Dover Publications, 1950.)

c. Mott, Sir Nevill F. *Elements of Wave Mechanics.* Cambridge, England: At the University Press, 1952. 156 pp.

d. Dirac, Paul A. M. *The Principles of Quantum Mechanics.* (Fourth Edition.) London: Oxford University Press, 1958. 312 pp.

e. Messiah, Albert. *Quantum Mechanics.* New York: Interscience Publishers, 1961-1962. 2 vols.

See also "Resource Letter QSL-1 on Quantum and Statistical Aspects of Light," by P. Carruthers. *American Journal of Physics,* 31: 321-325, May 1963.

Quantum theory is further applied to the study of matter in:

a. Bates, D. R., editor. *Quantum Theory.* New York: Academic Press, 1961-1962. 3 vols.

b. Slater, John C. *Quantum Theory of Atomic Structure.* New York: McGraw-Hill Book Company, 1960. 2 vols.

c. Slater, John C. *Quantum Theory of Molecules and Solids.* New York: McGraw-Hill Book Company, 1963-1966. 3 vols. A fourth volume is promised.

d. Kittel, Charles. *Quantum Theory of Solids.* New York: John Wiley and Sons, 1963. 435 pp.

e. Levine, Sumner N. *Quantum Physics of Electronics.* New York: The Macmillan Company, 1965. 301 pp.

f. Yariv, Amnon. *Quantum Electronics.* New York: John Wiley and Sons, 1967. 478 pp.

For philosophical and historical background, see:

a. Jammer, Max. *The Conceptual Development of Quantum Mechanics.* New York: McGraw-Hill Book Company, 1966. 399 pp.

b. Gamow, George. *Thirty Years That Shook Physics: The Story of Quantum Theory.* Garden City, N. Y.: Doubleday and Company, 1966. 224 pp.

c. Kuhn, Thomas S., *et al. Sources for History of Quantum Physics.* Philadelphia: American Philosophical Society, 1967. 176 pp. (Its *Memoirs*, Vol. 68.) Report on methods used and materials gathered by an archival project.

d. Waerden, Bartel L. van der, editor. *Sources of Quantum Mechanics.* Amsterdam: North-Holland Publishing Company, 1967. 442 pp. An historical introduction is followed by most important early papers, all in English text.

See also Sommerfeld's *Wave Mechanics;* Coulson's *Waves;* and Löwdin's *Quantum Theory . . . Tribute to John C. Slater.*

Relativity theory

Einstein's theory postulates that motion through free space is relative rather than absolute, and that the velocity of light is independent of source velocity. (The special theory of relativity shows that mass, i.e., inertia, varies with velocity.) Physicists may read the originator's own summary in:

Einstein, Albert. *The Meaning of Relativity.* (Fifth Edition.) Princeton, N. J.: Princeton University Press, 1955. 169 pp.

The beginner had better start with:

a. Lieber, Lillian R. *The Einstein Theory of Relativity.* New York: Rinehart and Company, 1945. 324 pp.

b. Bergmann, Peter G. *Introduction to the Theory of Relativity.* New York: Prentice-Hall, Inc., 1942. 287 pp.
This textbook covers mathematical and physical aspects, and is prefaced by a foreword written by Einstein. Its three parts are: Special theory of relativity; General theory of relativity; Unified field theories.

c. Barnett, Lincoln K. *Universe and Dr. Einstein.* (Second Edition.) New York: William Sloane Associates, 1957. 127 pp.
Includes foreword by Albert Einstein.

d. Schwartz, Jacob T. *Relativity in Illustrations,* designed and illustrated by Felix Cooper. New York: New York University Press, 1962. 117 pp.
Uses cartoons to answer the questions: What is time? What is space?

Other references on the general and special theories:

a. Adler, Ronald; Bazin, Maurice; and Schiffer, Menahem. *Introduction to General Relativity.* New York: McGraw-Hill Book Company, 1965. 451 pp.

b. Synge, John L. *Relativity: The General Theory.* New York: Interscience Publishers, 1960. 505 pp.

c. Synge, John L. *Relativity: The Special Theory.* (Second Edition.) New York: Interscience Publishers, 1965. 459 pp.

d. Rindler, W. *Special Relativity.* (Second Edition.) New York: John Wiley and Sons, 1966. 196 pp.

See also "Resource Letter SRT-1 on Special Relativity Theory," by Gerald Holton. *American Journal of Physics,* 30: 462-469, June 1962.

For the necessary mathematics, one may turn to:

Rainich, G. Y. *Mathematics of Relativity.* New York: John Wiley and Sons, 1950. 173 pp.

A comprehensive bibliography is available:

Lecat, Maurice. *Bibliographie de la Relativité.* Bruxelles: Maurice Lamertin, 1924. 290: 47 pp.
Arrangement is alphabetical by author, with each item numbered in square brackets. See "Table des Matières" (p. ix) for various component lists and index approaches, e.g., by periodical in which article had appeared; chronological history of articles by their dates (from 1728 to 1924).

For Einstein's writings there is:

Boni, Nell, *et al.* *A Biographical Checklist and Index to the Published Writings of Albert Einstein.* Paterson, N. J.: Pageant Books, 1960. 84 pp.

Summary

While mathematics furnishes many necessary tools to science, it must not presume to overshadow physical meanings and concepts. Fletcher, *et al.*, have listed mathematical tables of all kinds. Compendia of physical constants, essentially numerical, also vary widely from the Childs booklet to the monumental Landolt-Börnstein work. Applied mathematics has been discussed from the standpoints of technique in elementary problem solving as well as formulation of the subject content of advanced theoretical physics, notably quantum mechanics and relativity.

CHAPTER VII

EDUCATIONAL APPROACH

Study, Teaching, and Educational Research

The educational approach signifies desire to consult sources of information on the theory and practice of physics teaching, embracing a wide range of material from research to popularized treatments of the subject content of physics, which contribute to general education if soundly conceived.

Overview

The old psychology of physics teaching placed too much emphasis upon memorization of miscellaneous subject material. Whether the material figured importantly in life was unimportant. Interrelationships among diverse items were not brought out clearly.

Today the needs and interests of the learner are prime criteria. Whatever will help students meet life's exigencies is included in curricula, rather than masses of accumulated facts for their own sake. Clear constructive thinking and purposeful action are to be cultivated. All knowledge is to be integrated by delineating significant interrelationships and arriving at conclusive generalizations:

> To summarize then, the developers of the new curricula have attempted to cause the student to discover for himself major science generalizations. The authors have chosen to ignore the fact that this approach leaves large gaps in the amount of scientific knowledge which a given student acquires. It is assumed that a science cannot be "covered" in one semester, but the student provided with certain basic intellectual tools for attacking problems and with the basic scientific generalizations which underlie most of technology is likely to be able to adapt to a variety of situations. Further, it is hoped that the student will become scientifically literate in that he will have a better understanding of how and why scientists approach problems, and will therefore be less likely to be guilty of the kind of science ignorance which is so widespread today.[1]

The various books on the study and teaching of physical science mentioned later in this chapter embrace this enlightened viewpoint, including some of less recent publication date, which were evidently forerunners of the new era.

---

1. J. S. Marshall and E. Burkman, *Current Trends in Science Education*, p. 10. New York: Center for Applied Research in Education, Inc., 1966.

Early period.

McCarthy has written an interesting description [2] of the beginnings of physics teaching in American colleges. Conditions over a century later (1880) are depicted in the following governmental circulars:

a. Clarke, Frank W. *A Report on the Teaching of Chemistry and Physics in the United States.* Washington, D. C.: Government Printing Office, 1881. 219 pp. (U. S. Bureau of Education, Circular of Information No. 6, 1880.)
Activities of particular institutions are described, and early textbooks are listed (pp. 157-166).

b. Wead, Charles K. *Aims and Methods of the Teaching of Physics.* Washington, D. C.: Government Printing Office, 1884. 158 pp. (U. S. Bureau of Education, Circular of Information No. 7, 1884.)
This presents well-arranged material on physics teaching practices at all levels, collected from a wide range of sources, and developing an interesting picture of the period.

Transition.

The following surveys reveal science teaching to be a mixture of old and new practices:

a. Beauchamp, Wilbur L. *Instruction in Science.* Washington, D. C.: Government Printing Office, 1933. 63 pp. (U. S. Office of Education Bulletin 1932, No. 17, Monograph No. 22.)
Secondary school practices are extensively surveyed, showing trends towards better educational ideas and methods, such as development of major concepts, project method, visual aids, etc.

b. Underhill, Orra E. *The Origins and Development of Elementary School Science.* Chicago: Scott, Foresman and Company, 1941. 347 pp.
Methods and textbooks are surveyed from 1750 to 1939.

c. Powers, S. Ralph. "Physical Sciences in Our Secondary Schools." *American Journal of Physics,* 27: 419-423, September 1959.
Historical account of trends from the nineteenth century to 1959.

Recent period.

Since 1959 there has been fruitful activity on the part of physics teachers and their organizations toward improved teaching at both secondary and college levels, surveyed in:

---

2. J. J. McCarthy, "Physics in American Colleges Before 1750." *American Physics Teacher,* 7: 100-104, April 1939.

a. *Science Education in the Schools of the United States:* Report of the National Science Foundation to . . . Congress. Washington, D. C.: Government Printing Office, 1965. 125 pp.
Sections: 1, Science education through 1950; 2, Revolutions in science education: the last decade; 3, 4, Science education in the 1960's (Course content; Teacher education); and 5, Problems and issues. Includes a brief history of the PSSC (Physical Science Study Committee): pp. 67-70.

b. Marshall, James S., and Burkman, Ernest. *Current Trends in Science Education.* New York: Center for Applied Research in Education, Inc., 1966. 115 pp.
Especially Chap. 2: The transformation of high school physics—A case study, pp. 11-28.

As reported by Gerald Holton (*American Journal of Physics*, 28: 568-578, September 1960), a preliminary conference on college physics outlined objectives and recommended that the American Association of Physics Teachers establish a commission to implement them. This Commission on College Physics has made three progress reports [3] on its national program to date. Among its accomplishments is the paper-back series of "Momentum Books" which give lucid and accurate topical presentations, as listed in the D. Van Nostrand Company catalog.

At the secondary level, the Physical Science Study Committee (PSSC) was organized at M.I.T. in 1956, with representation from universities, high schools, and industry. It introduced a more meaningful and coherent course of study, described in:

a. Finlay, Gilbert C. "Secondary School Physics: The Physical Science Study Committee." *American Journal of Physics*, 28: 286-293, March 1960.

b. Arons, A. B. "The New High School Physics Course." *Physics Today*, 13 (No. 6): 20-25, June 1960.
Symposium held at the joint meeting of the American Association of Physics Teachers and the American Physical Society.

The PSSC also has a paper-back "Science Study" series in the Doubleday Anchor Book catalog, and a comprehensive film program through Educational Services, Inc., Watertown, Mass.

For progress and trends in introductory physics education, see *Physics Today*, 20 (No. 3): 25-79, March 1967. *Pre-College:* 1, The

---

3. *American Journal of Physics*, 30: 665-686, October 1962; 32: 398-427, June 1964; and 34: 833-861, September 1966.

PSSC course; 2, Harvard project physics; 3, Engineering concepts; and 4, The Nuffield project. *College:* 1, Teaching from Feynman; 2, The Berkeley course; 3, The new M.I.T. course; 4, Baccalaureate science; 5, Physical science for nonscientists; 6, Commission on college physics; and 7, Too far, too fast? Also covered are enrollment trends.

See also Arnold A. Strassenburg's "American Institute of Physics Programs in Education—Present and Future." (*American Journal of Physics,* 35: 797-807, September 1967.)

Study and Teaching

Sound teaching principles and techniques may be derived from books on science teaching in general, as well as those centered exclusively upon physics.

Physics teaching.

Current ideas on better physics teaching may be gleaned from:

a. Warren, John W. *The Teaching of Physics.* London: Butterworths Scientific Publications, 1965. 130 pp.

b. American Institute of Physics. *Physics in Your High School; A Handbook for the Improvement of Physics Courses.* New York: McGraw-Hill Book Company, 1960. 136 pp.
Authors: W. C. Kelly and Thomas D. Miner. Goals are stated for school board members with respect to course content, teaching, materials, etc.

c. United Nations Educational, Scientific and Cultural Organization. *A Survey of the Teaching of Physics at Universities.* New York: UNESCO, 1966. 396 pp.
Teaching procedures, materials, problems, etc., are surveyed in European countries and our own, yielding an international perspective on improved instruction at all levels.

d. International Conference on Physics Education, Paris 1960. *International Education in Physics,* edited by Sanborn C. Brown and Norman Clarke. Cambridge, Mass.: The M.I.T. Press, 1961. 191 pp.
Better physics teaching is of world-wide concern. *See also* F. W. Sears' summary (*American Journal of Physics,* 29: 151-160, March 1961).

e. International Conference on Physics in General Education, Rio de Janeiro 1963. *Why Teach Physics?* Edited by Sanborn C. Brown, Norman Clarke, and Jayme Tiomno. Cambridge, Mass.: The M.I.T. Press, 1964. 97 pp.
Need for progress in many countries.

f. International Conference on the Education of Professional Physicists, London 1965. *The Education of a Physicist,* edited by Sanborn C. Brown and Norman Clarke. Cambridge, Mass.: The M.I.T. Press, 1966. 185 pp.

Although lacking in desired recency, the following may be helpful on the teaching of physics *per se:*

a. Rusk, Rogers D. *How to Teach Physics.* Philadelphia: J. B. Lippincott Company, 1923. 186 pp.
The author maintains that physics teaching should be integrated with life needs, displaying an enlightened viewpoint for the period.

b. Mann, C. Riborg. *The Teaching of Physics for Purposes of General Education.* New York: The Macmillan Company, 1922. 304 pp.
Humanizing of physics is advocated. Historical aspects are well presented.

A few special-subject treatments follow:

a. American Physical Society. *The Teaching of Physics, with Special Reference to the Teaching of Physics to Students of Engineering.* New York: The Society, 1922. 55 pp.
Its contents include: Purpose of physics in an engineering college; Methods of instruction in physics (lecture, recitation and laboratory); Defects of present methods as seen by others; Co-ordination of physics and other subjects in engineering colleges.

b. Mathematical Association. *The Teaching of Mechanics in Schools.* London: G. Bell and Sons, 1930. 84 pp.
Concrete suggestions are given for relating mechanics to everyday life, rather than general philosophical aspects.

c. Physical Society. *Teaching of Geometrical Optics.* London: The Society, 1934. 86 pp.

d. American Association of Physics Teachers. "The Teaching of Geometrical Optics." *American Physics Teacher,* 6: 78-82, April 1938.

e. American Association of Physics Teachers. "The Teaching of Electricity and Magnetism at the College Level." *American Journal of Physics,* 18: 1-25; 69-88, January and February 1950.

Science teaching.

Interesting overall surveys of science education are provided by:

a. Nedelsky, Leo. *Science Teaching and Testing.* New York: Harcourt, Brace and World, 1965. 368 pp.
Newer ideas in physics teaching are presented, with evaluative tests.

b. Haun, Robert R., editor. *Science in General Education.* Dubuque, Iowa: William C. Brown Company, 1960. 291 pp.
Reports from many colleges on their courses and philosophy.

c. Estrin, Herman A., editor. *Higher Education in Engineering and Science.* New York: McGraw-Hill Book Company, 1963. 548 pp.
Authoritative articles on good teaching techniques and approaches.

d. Brandwein, Paul F.; Watson, Fletcher G.; and Blackwood, Paul E. *Teaching High School Science.* New York: Harcourt, Brace and World, 1958. 568 pp.

e. Noll, Victor H. *The Teaching of Science in Elementary and Secondary Schools.* New York: Longmans, Green and Company, 1942. 238 pp.
This is a comprehensive digest of thought and published writings on science teaching. It has lengthy bibliographies on such topics as objectives, scientific attitude, laboratory vs. lecture demonstration, etc. In the appendix, pp. 222-227, are listed the best objective tests in physics.

f. Richardson, John S. *Science Teaching in Secondary Schools.* Englewood Cliffs, N. J.: Prentice-Hall, Inc., 1957. 385 pp.

g. Thurber, Walter A., and Collette, Alfred T. *Teaching Science in Today's Secondary Schools.* (Second Edition.) Boston: Allyn and Bacon, 1964. 701 pp.

h. Washton, Nathan S. *Science Teaching in the Secondary School.* New York: Harper and Row, 1961. 328 pp.

See also selected yearbooks of the National Society for the Study of Education, e.g., 31st, Part 1: *A Program for Teaching Science* (1932); 46th, Part 1: *Science Education in American Schools* (1947); and 59th, Part 1: *Rethinking Science Education* (1960).

Background materials are featured in:

a. Richardson, John S., and Cahoon, G. P. *Methods and Materials for Teaching General and Physical Science.* New York: McGraw-Hill Book Company, 1951. 485 pp.
The authors show how to enrich high school and introductory college science courses, with emphasis upon physics. Demonstrations, experiments and projects are presented, and sources indicated.

b. Heiss, Elwood D.; Obourn, E. S.; and Hoffman, C. W. *Modern Science Teaching.* New York: The Macmillan Company, 1950. 462 pp.
This is a revision of the authors' earlier *Modern Methods and Materials for Teaching Science,* and is intended as both a textbook in science education and a source book of equipment and sensory aids.

c. Woodring, Maxie N., *et al.* *Enriched Teaching of Science in High School.* (Second Edition.) New York: Bureau of Publications, Teachers College, Columbia University, 1941. 402 pp.

d. Joseph, Alexander, *et al.* *Teaching High School Science: A Sourcebook for the Physical Sciences.* New York: Harcourt, Brace and World, 1961. 674 pp.

e. Spielman, Harold S. *Electronics Source Book for Teachers.* New York: Hayden Book Company, 1965. 3 vols.

f. Meitner, John G., editor. *Astronautics for Science Teachers.* New York: John Wiley and Sons, 1965. 381 pp.
Resources for teaching astronautics: pp. 355-367.

g. *700 Science Experiments for Everyone;* foreword by Gerald Wendt; originally published as UNESCO *Source Book for Science Teaching.* (Revised and Enlarged Edition.) Garden City, N. Y.: Doubleday and Company, 1964. 250 pp.

"Resource Letters" have been appearing in the *American Journal of Physics* since the March 1962 issue. The first twenty-one are available from the American Institute of Physics in collective form as *Resource Letters, Book One,* ending with "Resource Letter PB-1 on Physics and Biology" in the February 1966 issue of the journal. Other collective *Reprint Books* include a particular resource letter and actual reprints of selected items mentioned therein. The intent is "to guide college physicists to some of the literature and other teaching aids that may help them improve course contents in specified fields of physics." The project, started by the Commission on College Physics, is now directed by the American Association of Physics Teachers.

Because the Resource Letters include journal articles, etc., beyond the scope of this *Guide,* their citations are included under subject insofar as publication date permitted. Any later ones may be found, of course, in the *American Journal of Physics.*

See also the lecture demonstrations, audio-visual materials, etc., mentioned under Presentational Approach in Chapter IX, especially *Educational Media Index.*

Testing devices are discussed in books by Nedelsky, Noll, Heiss, *et al.,* above, and in:

a. Kruglak, Haym. "Achievement Testing." *American Journal of Physics,* 33: 255-263, April 1965.
This is Resource Letter AT-1 of an AAPT series, and furnishes a guide to testing literature at the college level.

b. Kruglak, Haym, and Wall, C. N. *Laboratory Performance Tests*

for General Physics. Kalamazoo, Mich.: Western Michigan University, 1959. 165 pp.
Grading students on their laboratory technique rather than the usual report.

c. Buros, Oscar K., editor. *The Sixth Mental Measurements Yearbook.* Highland Park, N. J.: Gryphon Press, 1965. 1714 pp.
Physics tests are described and reviewed on pp. 1194-1201.

d. Ferris, Frederick L., Jr. "Testing for Physics Achievement." *American Journal of Physics,* 28: 269-278, March 1960.
Based on Physical Science Study Committee course.

e. Hedges, William D. *Testing and Evaluation for the Sciences in the Secondary School.* Belmont, Cal.: Wadsworth Publishing Co., 1966. 248 pp.
Helpful comments on the mechanics of testing.

f. "Measuring the Results of Instruction in College Physics." (A summary report on the National College Physics Testing Program, 1933-1939.) *American Journal of Physics,* 8: 173-181, June 1940.
Four previous reports are mentioned in a footnote (p. 174).

Periodicals.

A discussion of the periodicals of interest to the physicist was given in Chapter II of this guide. From the educational viewpoint, the most fruitful source of articles on physics teaching at the college level is the *American Journal of Physics* (formerly the *American Physics Teacher*). For elementary and secondary education, the corresponding source is *School Science and Mathematics.* New high-school journals have appeared: *The Physics Teacher* (American Association of Physics Teachers); and *Journal of Physics Education* (Institute of Physics and the Physical Society). The American Institute of Physics publishes an international newsletter entitled *Physics Education.* Indexing media for contents of journals are *Applied Science and Technology Index* and *Education Index.* (Unfortunately, *Physics Abstracts* does not bother to include in its indexing any general articles, i.e., articles on other than the topics of physics *per se.*) For further information on serial indexes, refer to Chapter II of this guide, and to Burke's, cited below under Educational Research.

Bibliographies.

Extensive references on science teaching may be derived from:

a. Glenn, Earl R., and Walker, Josephine. *Bibliography of Science Teaching in Secondary Schools.* Washington, D. C.: Government

Printing Office, 1925. 161 pp. (U. S. Office of Education Bulletin 1925, No. 13.)
This is valuable because it analyzes selected periodicals prior to the beginning of *Education Index* in 1929. Physics is covered on pp. 132-160.

b. Monroe, Walter S., and Shores, Louis. *Bibliographies and Summaries in Education to July, 1935.* New York: The H. W. Wilson Company, 1936. 470 pp.

c. Hollingsworth, J. R. "Abridged Bibliography of Studies Pertaining to Science Teaching." *American Journal of Physics,* 9: 297-303, October 1941.
A numbering scheme groups the references under sixty topics, e.g., Rote Learning; Lecture Method.

Current bibliographies are offered by the *Bibliographic Index,* previously noted. Under the headings "Physics—Study and Teaching" and "Science—Study and Teaching" may be found relevant books in the *Cumulative Book Index.*

Study.

Shifting emphasis from teacher to pupil, one finds the complementary aspect of effective study helpfully presented in these manuals:

a. Chapman, Seville. *How to Study Physics.* (Second Edition.) Reading, Mass.: Addison-Wesley Publishing Company, 1949. 34 pp.

b. Sanford, Fernando. *How to Study; Illustrated through Physics.* New York: The Macmillan Company, 1922. 56 pp.

c. Crawford, Claude C. *Studying the Major Subjects.* Los Angeles, Cal.: University of Southern California, 1930. 384 pp.
Physics and chemistry are jointly treated, pp. 129-159.

Reading skill is basic to effective study, and may be stimulated by:

a. Howland, Hazel P.; Jarvie, L. L.; and Smith, L. F. *How to Read in Science and Technology.* New York: Harper and Brothers, 1943. 264 pp.
Sections include: Reading for details; Reading for main ideas; Reading to understand principles; Reading to follow directions; Reading to solve a problem; Reading to understand and interpret graphical materials. Selected passages are presented, followed by exercises thereon.

b. Bloomer, Richard H. *Reading Comprehension for Scientists.* Springfield, Ill.: Charles C Thomas, 1963. 213 pp.
Aims to alleviate the scientist's reading problem through exercises in the comprehension of passages lacking many key words.

Additional help in forming desirable study habits may be obtained from Chapter I, "The Principles of Reading and Study," in Parke. *See also* Dadourian's *How to Study; How to Solve,* and pp. xi-xv of Lindsay's *Student's Handbook.*

For formal study on a higher level consult:

a. American Council on Education. *A Guide to Graduate Study; Programs Leading to the Ph.D. Degree,* edited by Jane Graham. (Third Edition.) Washington, D. C.: The Council, 1965. 609 pp. Programs, requirements, fees, etc.

b. Institute of International Education. *Handbooks on International Study.* (Fourth Edition.) New York: The Institute, 1965. 2 vols.
The first volume is for Americans studying abroad, and the second for foreign nationals taking courses here.

See also index under Fellowships for *Study Abroad.*

Popularization.

It is salutary for those associated with educational institutions to be reminded occasionally that general education is not confined to the classroom. Self-teaching on an informal and entirely voluntary basis is open to all who possess the ambition and persistence to learn. Dingle [4] even goes further in stating that "when the terms employed are clearly defined and the sentences used are unambiguous, the mental grasp of the educated layman is not inferior to that of the scientist."

Suggested books for introductory purposes may be found in:

Deason, Hilary J. *The AAAS Science Book List for Young Adults.* Washington, D. C.: American Association for the Advancement of Science, 1964. 250 pp.

Only a few attractive subject matter presentations need be cited among many:

a. Jones, Gwyn O.; Rotblat, J.; and Whitrow, G. J. *Atoms and the Universe.* (Second Edition.) New York: Charles Scribner's Sons, 1963. 277 pp.

b. Beiser, Germaine, and Beiser, Arthur. *Physics for Everybody.* New York: E. P. Dutton and Company, 1956. 191 pp.

c. Rothman, Milton A. *The Laws of Physics.* New York: Basic Books, Inc., 1963. 254 pp.
Some chapters: What is a law of nature?; Conservation of energy; Laws of motion; Forces and fields; Elementary particles.

---

4. Herbert Dingle, *Modern Astrophysics,* p. x. (Second Edition.) New York: The Macmillan Company, 1927.

d. Newlon, Clarke. *1001 Questions Answered About Space.* New York: Dodd, Mead and Company, 1962. 355 pp.

e. Freeman, Ira M. *Physics Made Simple.* (Revised Edition.) Garden City, N. Y.: Doubleday and Company, 1965. 192 pp.

f. Gamow, George. *Matter, Earth and Sky.* (Second Edition.) Englewood Cliffs, N. J.: Prentice-Hall, Inc., 1965. 624 pp.
An interesting unified treatment requiring concentrated thought.

g. Rogers, Eric M. *Physics for the Inquiring Mind; The Methods, Nature and Philosophy of Physical Science.* Princeton, N. J.: Princeton University Press, 1960. 778 pp.
Pleasurable, informal, mature.

Popular fallacies are dispelled in the following compilations:

a. Ackermann, Alfred S. E. *Popular Fallacies; A Book of Common Errors Explained and Corrected.* (Fourth Edition.) London: Old Westminster Press, 1950. 843 pp.
For example, one would find in this compilation such false physics notions as warm water freezing faster than cold on wintry pavements.

b. Hampson, W. *Paradoxes of Nature and Science.* New York: E. P. Dutton and Company, 1907. 304 pp.

c. Hering, Daniel W. *Foibles and Fallacies of Science.* New York: D. Van Nostrand Company, 1924. 294 pp.

d. Phin, John. *The Seven Follies of Science.* London: Arnold Constable and Company, 1906. 178 pp.
Phin relates the most famous scientific impossibilities, notably perpetual motion.

e. Dircks, Henry. *Perpetuum Mobile; Or, Search for Self-Motive Power, during the 17th, 18th, and 19th Centuries.* London: E. and F. N. Spon, 1861. 558 pp.

Scientific analyses of psychic phenomena are undertaken in:

a. Still, Alfred. *Borderlands of Science.* New York: Philosophical Library, 1950. 424 pp.
The author states (p. 1):

> This book is not a history of science; neither is it a history of magic; it is an attempt to evaluate the influence on civilization of both science and superstition—knowledge and belief, and to consider critically those "borderland" phenomena which the scientist rarely investigates notwithstanding that they occur in the natural world which he shares with the unreasoning multitude.

Some of the chapters are: Science and the scientist; Magic and the mystic; Witchcraft and the new science; The divining rod; Levitation; Hypnotism and clairvoyance; Telepathy.

b. Tromp, Solco W. *Psychical Physics.* New York: Elsevier Publishing Company, 1949. 534 pp.

As a geologist, the author encountered many dowsers—people claiming the power to use divining rods to discover water or underground deposits. He examines evidence of the influence of electromagnetic fields on living organisms, with respect to divining and similar phenomena.

Pseudo-science and quackery are exposed in:

Gardner, Martin. *Fads and Fallacies in the Name of Science.* (Second Edition.) New York: Dover Publications, 1957. 363 pp.

(For students' misconceptions in physics, see lists[5] by Perkins, *et al.*)

Educational Research

Before special research is attempted in science education, procedural guidance in general educational research should be obtained from the following:

a. Burke, Arvid J., and Burke, Mary A. *Documentation in Education:* Revision of Carter Alexander and Arvid J. Burke: *How to Locate Educational Information and Data,* Fourth Edition, Revised. New York: Teachers College Press, 1967. 413 pp.

This is an outstanding example of the art of guide compilation. It surveys all types of library materials, and outlines most efficient utilization. "The transfer [to A. J. Burke] gives me the pride and happiness of a senior partner when the junior he developed to succeed him takes over fully," wrote Dr. Alexander in the foreword to this "fifth edition" on August 17, 1965. On August 24th he passed away.

b. Good, Carter V. *Essentials of Educational Research; Methodology and Design.* New York: Appleton-Century-Crofts, Inc., 1966. 429 pp.

c. Travers, Robert M. W. *An Introduction to Educational Research.* (Second Edition.) New York: The Macmillan Company, 1964. 581 pp.

d. *Handbook of Research on Teaching;* A project of the American Educational Research Association; edited by N. L. Gage. Chicago: Rand McNally and Company, 1963. 1218 pp.

"Research on Teaching Science," by F. G. Watson: pp. 1031-1059.

---

5. *American Journal of Physics,* 11: 101-102; 163-165; 227-228, April-August 1943.

e. *Encyclopedia of Educational Research,* edited by Chester W. Harris. (Third Edition.) New York: The Macmillan Company, 1960. 1564 pp.
Earlier editions by Walter S. Monroe.

General lists.

The general field of educational research is spanned chronologically by the following:

a. Monroe, Walter S. *Ten Years of Educational Research, 1918-27.* Urbana, Ill.: University of Illinois, 1928. 377 pp. (University of Illinois, Bureau of Educational Research, Bulletin No. 42, August 1928.)

b. U. S. Office of Education. *Bibliography of Research Studies in Education, 1926-27—1939-40.* Washington, D. C.: Government Printing Office, 1929-1942. Vols. 1-14. (Its *Bulletin* 1928-1941.)
See especially the "Chemistry and Physics" sections.

c. Besides annual lists in the *Phi Delta Kappan,* decennial subject indexes of *Research Studies in Education* have been compiled by the society for 1941-1951; 1953-1963. (Separate index available for 1952.)

Current educational research studies may be found in the *Review of Educational Research,* the *Journal of Educational Research,* etc.; in a monthly catalog, *Research in Education,* issued by the U. S. Office of Education; and in *Education Index* (which has concentrated more on journal content since 1961).

See also the section Dissertations of this *Guide.*

Science lists.

As a review of published research in the teaching of science from early in the century to the year 1937, the following series of digests is helpful, not only for listings of research studies but especially for keen analyses under headings of problem, method, findings, etc. Only three have ever appeared:

Curtis, Francis D. *A Digest of Investigations in the Teaching of Science.* Philadelphia: Blakiston and Company, 1926. 341 pp.
Research investigations published prior to 1925 are included.

Curtis, Francis D. *Second Digest . . .* Philadelphia: Blakiston and Company, 1931. 424 pp. (Covering 1925 through 1930.)

Curtis, Francis D. *Third Digest . . .* Philadelphia: Blakiston and Company, 1939. 419 pp. (Covering 1931 through 1937.)
These compilations, which include supplementary bibliographies as

well as outlines of selected researches, are useful to research workers, science teachers and school administrators. They have been continued by Robert Boenig's *Study of the Research Done in Science Education During the Years 1938-1947*, an unpublished Ed.D. report reproduced on microfilm (Teachers College, Columbia University, 1963).

The U. S. Office of Education reviews *Research in the Teaching of Science* from time to time, e.g., in its *Bulletin 1962, No. 3,* covering 1957-1959; and *1965, No. 10* for 1959-1961.

See also the quarterly *Journal of Research in Science Teaching*, published under the auspices of the National Association for Research in Science Teaching, and Association for the Education of Teachers in Science.

Summary

Physics teaching has gradually joined the trend away from mere rote-learning to real-life significance as determined by needs and interests. The various textbooks, surveys and current records cited provide helpful suggestions on how to teach. Conversely, how to study is explored in several booklets written from the learner's viewpoint. Popularization of physics content (without distortions and inaccuracies) is sound introductory procedure. For educational research, Burke's revision of Alexander's comprehensive guide covers existing sources of data and their efficient use, as steps towards still further knowledge and developments.

CHAPTER VIII

TERMINOLOGICAL APPROACH
Definitions and Translations

When one seeks definitions of terms, or assistance in translating them, he follows the terminological approach. Abbreviations and symbols are related entities.

Overview

"A definition of a term or quantity is an expression stated in simpler or more fundamental terms or quantities, that may replace the original term or quantity wherever used without loss or change of thought." [1]

Roller has prepared an interesting discussion of the idiosyncrasies of physical terms:

Roller, Duane.[2] *The Terminology of Physical Science.* Norman, Okla.: University of Oklahoma Press, 1929. 115 pp.
The chapters are: Physical terms and their definitions; Common prefixes and suffixes; Names of the chemical elements; Pronunciation of words used in science; Spelling of words used in science; Simpler standard abbreviations. Teachers and textbook writers are urged to avoid careless or ambiguous terminology which proves troublesome to students.

The structure of scientific terms, and their formation from roots, stems, prefixes and suffixes of Latin and Greek origin, will be clarified by:

a. Hough, John N. *Scientific Terminology.* New York: Rinehart and Company, 1953. 231 pp.

b. Brown, Roland W. *Composition of Scientific Words: A Manual of Methods and a Lexicon of Materials for the Practice of Logotechnics.* Washington, D. C.: The Author, 1954. 882 pp.

c. Flood, W. E. *Scientific Words, Their Structure and Meaning: An Explanatory Glossary of About 1150 Word-Elements (Roots,*

---

1. A. G. Worthing, "A Simple Test for the Preciseness of a Definition." *American Physics Teacher,* 6: 59-61, April 1938.
2. See also his "An Approach to the Study of Physical Terminology." *American Journal of Physics,* 15: 178-186, March-April 1947.

Prefixes, Suffixes) Which Enter into the Formation of Scientific Terms. New York: Duell, Sloan and Pearce, 1960. 220 pp.

Interrelated terms are charted in:

Engineers Joint Council. *Thesaurus of Engineering and Scientific Terms.* (Revised Edition.) New York: The Council, 1967. 800 pp. "Intended primarily to serve technical information personnel and editors as a vocabulary reference for indexing and retrieving technical information."

Every field of activity has its body of accepted meanings, even the art of bell-ringing, from which Taylor quotes an interesting passage:

> A plain course in any method may be extended into a touch or even a peal by calling a series of bobs or singles or both. When a bob is called, the bell that was about to make seconds runs out quick and lies behind for a whole pull. The bell that was about to dodge in 3-4 down hunts down instead and turns the treble from lead. . . . The beginner should familiarize himself with the composition of touches by repeatedly pricking changes until he can dependably bring the bells into rounds from any lead-end at which he may start.[3]

These directions seem simple and matter-of-fact to a bell ringer, because he is quite familiar with the terms. Definitions must thus be learned before one may proceed intelligently with practical or theoretical work of any kind, particularly in the exact sciences.

Definitions

For meanings of all but the most technical terms, the subject specialist should not disdain using the two principal unabridged general dictionaries, namely, *Funk and Wagnalls New Standard Dictionary* and *Webster's New International Dictionary.* The general encyclopedias will also prove helpful on many occasions.

Physics dictionaries.

Deriving obvious advantages from their subject concentration, dictionaries for the field of physics as a whole include the following assorted types:

a. Gray, Harold J., editor. *Dictionary of Physics.* New York: Longmans, Green and Company, 1958. 544 pp.
This is for classical rather than modern terms, and has bibliographical and biographical features.

b. *International Dictionary of Physics and Electronics,* edited by

---

3. L. W. Taylor, *Physics; The Pioneer Science,* p. 13. Boston: Houghton Mifflin Company, 1941.

Walter C. Michels. (Second Edition.) Princeton, N. J.: D. Van Nostrand Company, 1961. 1355 pp.
Emphasis is more modern than in Gray's dictionary.

c. Thewlis, J., editor. *Encyclopaedic Dictionary of Physics: General, Nuclear, Solid State, Molecular, Chemical, Metal and Vacuum Physics, Geophysics, Biophysics, and Related Subjects.* Oxford: Pergamon Press, 1961-1964. 9 vols.
Author and subject indexes are in Vol. 8, and multilingual equivalents in Vol. 9. This comprehensive and monumental work is useful in conjunction with Flügge's Handbuch. It will be updated by supplementary volumes.

d. Besançon, Robert M., editor. *Encyclopedia of Physics.* New York: Reinhold Publishing Corporation, 1966. 832 pp.
Articles go beyond mere definitions.

e. Weld, LeRoy Dougherty. *Glossary of Physics.* New York: McGraw-Hill Book Company, 1937. 255 pp.
"The sole purpose is to give information as to actual usage, and in such form as to be intelligible to students as well as to specialists." As terms are always more easily understood in context, passages containing them are cited. This dictionary is of convenient size for frequent use.

f. Franke, H. *Lexikon der Physik.* (Zweite Auflage.) Stuttgart: Franckh'sche Verlagshandlung, 1959. 2 vols.
A well-illustrated encyclopedic dictionary.

g. Westphal, Wilhelm H. *Physikalisches Wörterbuch.*[4] Berlin: Springer, 1952. 2 vols. in 1. (833, 795 pp.)

h. Glazebrook, Sir Richard. *A Dictionary of Applied Physics.* London: The Macmillan Company, 1922-1923. 5 vols.
This comprehensive encyclopedic dictionary is divided as follows: (Vol. 1) Mechanics, engineering, heat; (Vol. 2) Electricity; (Vol. 3) Meteorology, metrology and measuring apparatus; (Vol. 4) Light, sound, radiology; and (Vol. 5) Aeronautics, metallurgy. Each volume is arranged by broad subject, with cross-references from minor topics. The set is of historical interest for detailed expositions.

See also index under Handbooks.

Certain subdivisions of applied physics have their own compendia:

a. Markus, John. *Electronics and Nucleonics Dictionary.* (Third Edition.) New York: McGraw-Hill Book Company, 1966. 743 pp.

---

4. Supersedes A. Berliner and K. Scheel, *Physikalisches Handwörterbuch.* (Zweite Auflage.) Berlin: Springer, 1932. 1428 pp.

b. Sarbacher, Robert. *Encyclopedic Dictionary of Electronics and Nuclear Engineering.* Englewood Cliffs, N. J.: Prentice-Hall, Inc., 1959. 1417 pp.

c. Roget, Samuel R. *A Dictionary of Electrical Terms.* (Fourth Edition.) London: Sir Isaac Pitman and Sons, 1941. 432 pp.
This covers "the general science of electricity and magnetism, without proceeding too far into its ramifications in the directions of pure physics and chemistry, etc."

d. National Research Council. Conference on Glossary of Terms in Nuclear Science and Technology. *A Glossary of Terms in Nuclear Science and Technology.* New York: American Society of Mechanical Engineers, 1957. 188 pp.

e. Del Vecchio, Alfred. *Concise Dictionary of Atomics.* New York: Philosophical Library, 1964. 262 pp.
Simply-worded definitions for introductory purposes.

In connection with seminars for science writers,[5] the American Institute of Physics has issued brief *Glossaries of Terms Frequently Used in:* Accelerators; Acoustics; Biophysics; Cryogenics; High energy physics; Lasers; Optics and spectroscopy; Plasma physics; Solid state physics; Space physics; etc.

Named effects and laws may be identified in:

a. Hix, C. F., Jr., and Alley, R. P. *Physical Laws and Effects.* New York: John Wiley and Sons, 1958. 291 pp.

b. Ballentyne, D. W. G., and Walker, L. E. Q. *A Dictionary of Named Effects and Laws in Chemistry, Physics and Mathematics.* (Second Edition.) London: Chapman and Hall, Ltd., 1961. 234 pp.

Incidentally, a whole book sometimes appears on a particular effect:

a. Putley, E. H. *The Hall Effect and Related Phenomena.* Washington, D. C.: Butterworths Scientific Publications, 1960. 263 pp.

b. Gill, Thomas P. *The Doppler Effect.* London: Logos Press, 1965. 149 pp.

c. Frauenfelder, Hans. *The Mössbauer Effect.* New York: W. A. Benjamin, Inc., 1962. 336 pp. (Rudolf L. Mössbauer received a Nobel prize in 1961 for his discovery of the recoilless resonance absorption of gamma rays in atomic nuclei, with applications in precision measurements.)

d. Wertheim, Gunther K. *Mössbauer Effect; Principles and Applications.* New York: Academic Press, 1964. 116 pp.

For research results and bibliography see the *Mössbauer Effect Data*

---

5. See *Physics Today,* 19 (No. 1): 9-10, January 1966.

Index, 1958-1965, compiled by Arthur H. Muir, *et al.* (Interscience, 1967.)

See also "Resource Letter ME-1 on the Mössbauer Effect," by G. K. Wertheim. *American Journal of Physics*, 31: 1-6, January 1963.

Further sources of definitions of physics terms are Lindsay's handbook, and the USA Standards Institute's terminologies. For working definitions one may often rely on certain physics textbooks and handbooks.

Science dictionaries.

For broader coverage of science in general, one has the following encyclopedias and dictionaries:

a. *Van Nostrand's Scientific Encyclopedia.* (Third Edition.) Princeton, N. J.: D. Van Nostrand Company, 1958. 1839 pp.
This useful large-sized volume has articles of varying length arranged in one alphabet; additional articles may be found for all terms that appear in boldface type. In this revision much new material has been added.

b. *McGraw-Hill Encyclopedia of Science and Technology: An International Reference Work.* (Revised Edition.) New York: McGraw-Hill Book Company, 1966. 15 vols.
Very extensive coverage is furnished for intelligent laymen, college students, *et al.* Vol. 15 is a detailed index. The set is updated by *McGraw-Hill Yearbooks of Science and Technology* beginning with 1961 (published 1962). There is a 32-page teacher's guide, published in 1961. *See also* the *McGraw-Hill Basic Bibliography of Science and Technology*, and *McGraw-Hill Modern Men of Science.*

c. *Harper Encyclopedia of Science*, edited by James R. Newman. New York: Harper and Row, 1963. 4 vols.

d. *Chambers's Technical Dictionary*, edited by C. F. Tweney and L. E. C. Hughes. (Third Edition.) Edinburgh and London: W. and R. Chambers, Ltd.; New York: The Macmillan Company, 1958. 1028 pp.
About 60,000 terms of science and technology are defined.

e. Uvarov, E. B., and Chapman, D. R. *Dictionary of Science.* (Fourth Edition.) Baltimore, Md.: Penguin Books, Inc., 1960. 336 pp.

Abbreviations and symbols.

General and special compendia:

a. Zimmerman, O. T., and Lavine, Irvin. *Scientific and Technical*

Abbreviations, Signs and Symbols. (Second Edition.) Dover, N. H.: Industrial Research Service, 1949. 541 pp.

b. Arnell, Alvin. *Standard Graphical Symbols: A Comprehensive Guide for Use in Industry, Engineering and Science.* New York: McGraw-Hill Book Company, 1963. 534 pp.

c. Witty, M. B., *et al. Dictionary of Electronics Abbreviations.* New York: SETI (Scientist and Engineer Technological Institute) Publishers, 1961. 250 pp.

In addition, twelve attractively designed dictionaries of abbreviations, signs and symbols are being published in the Odyssey Scientific Library under the editorship of David D. Polon, *et al.* There are volumes for electronics; physics and mathematics; electricity; nuclear science; etc. Each volume is conveniently identified by its title initials, e.g., DPMA is boldly stamped on the *Dictionary of Physics & Mathematics Abbreviations, Signs & Symbols.* New York: Odyssey Press, 1966. 333 pp.

Committee reports include:

a. International Union of Pure and Applied Physics. "Symbols, Units and Nomenclature in Physics." *Physics Today,* 15 (No. 6): 20-30, June 1962.

b. "American Standard Letter Symbols for Physics." Final Report (No. 4) of the Committee on Letter Symbols. *American Journal of Physics,* 16: 164-179, March 1948.

Physics symbols and a discussion of the principles of selection.[6] These symbols also appear in the Chemical Rubber Publishing Company's handbook, and in a separate *USA Standard* Z10.6-1948.

c. Royal Society of London. Symbols Committee. *Symbols, Signs and Abbreviations Recommended for British Scientific Publications.* London: The Royal Society, 1951. 19 pp.

The United States of America Standards Institute and the British Standards Institution have promulgated standard periodical title abbreviations.

Translations

The NBS Clearinghouse for Federal Scientific and Technical Information, and the SLA Translations Center at the John Crerar Library in Chicago, collect translated technical material from many sources,

---

6. See also D. Roller, "A Proposed Procedure for Selecting and Using Symbols for Physical Units." *American Journal of Physics,* 21: 293-296, April 1953.

especially Russian, in order that our scientists may keep abreast of foreign developments. At the end of 1967, the Clearinghouse discontinued its *Technical Translations*, which had been the successor to *Translation Monthly* (SLA 1955-1958) and an earlier LC bibliography. Henceforth all U. S. Government-sponsored translatoins will be announced in *U. S. Government R & D Reports.* Other translations will appear in the *Translations Register-Index,* published since mid-1967 by the SLA Translations Center, and in the *ETC Quarterly Index* (European Translations Centre, Delfte, The Netherlands).

Information and directory media:

a. United Nations Educational, Scientific and Cultural Organization. *Scientific and Technical Translating; and Other Aspects of the Language Problem.* Paris: UNESCO, 1957. 282 pp.
Discusses quantitative and qualitative aspects of translating; methods and organizations; language study; etc. Bibliography of books for technical language study: pp. 266-274.

b. Kaiser, Frances E., editor. *Translators and Translations: Services and Sources in Science and Technology.* (Second Edition.) New York: Special Libraries Association, 1965. 214 pp.
Includes directory of translators; pools and information sources for translations; and bibliographies of translations (in full chronological detail, with subject indexes by item number).

c. Bower, William W. *International Manual of Linguists and Translators.* New York: The Scarecrow Press, 1959. 451 pp. *1st Supplement,* 1961. 450 pp.
General information and specific directory. List of dictionaries, manuals, grammars, etc.: pp. 105-295.

The American Institute of Physics publishes English translations of many Russian journals.

For news of translated books one may consult the new series of volumes of *Index Translationum,* an international bibliography of translations that resumed publication under UNESCO auspices in 1949. It lists translations published in the languages of many countries (including the United States), under subheadings such as "Natural and Exact Sciences."

Armstrong has a word of caution:

> One must inquire, then, whether the present great interest in translated books is really rational. Certainly a translated book is at some disadvantage: the author and the translator can hardly help being occasionally at cross-purposes; the examples are likely to be a bit unfamiliar, and any references for further reading may not be very accessible. It would seem to follow, then, that books of special excellence should be translated, but

for the most part home-grown ones are to be preferred. If this principle has not been followed recently, this should be considered as a fad of the times.[7]

Several bibliographies of foreign-language and English dictionaries are available:

a. Collison, Robert L. *Dictionaries of Foreign Languages; A Bibliographical Guide to the General and Technical Dictionaries of the Chief Foreign Languages.* London: Hafner Publishing Company, 1955. 210 pp.

b. United States. Library of Congress. General Reference and Bibliography Division. *Foreign Language—English Dictionaries.* (Second Edition.) Washington, D. C.: Library of Congress, 1955. 2 vols.

Vol. 1 lists special subject dictionaries, with emphasis upon science and technology; Vol. 2 is for general ones.

c. Marton, Tibor W. *Foreign-Language and English Dictionaries in the Physical Sciences and Engineering; A Selected Bibliography 1952-63.* Washington, D. C.: Government Printing Office, 1964. 189 pp. (U. S. National Bureau of Standards Miscellaneous Publication No. 258.)

d. United Nations Educational, Scientific and Cultural Organization. *Bibliography of Monolingual Scientific and Technical Glossaries,* by Eugen Wüster. New York: Columbia University Press, 1955-1959. 2 vols.

Vol. 1: National standards; Vol. 2: Miscellaneous sources.

e. United Nations Educational, Scientific and Cultural Organization. *Bibliography of Interlingual Scientific and Technical Dictionaries.* (Fourth Edition.) New York: Columbia University Press, 1961. 236 pp. *Supplement,* 1964. 83 pp.

f. Walford, Albert J. *A Guide to Foreign Language Grammars and Dictionaries.* London: The Library Association, 1964. 132 pp.

When language equivalents rather than definitions of words are sought, one turns to bi-lingual or multi-lingual dictionaries.

General foreign dictionaries.

A few well-known examples follow:

a. *Langenscheidt's Concise German Dictionary.* New York: Barnes and Noble, Inc., 1964. 2 vols. in 1. Volumes are 1, English-German

7. H. L. Armstrong, in review. *American Journal of Physics,* 34: 827, September 1966.

(Second Edition); and 2, German-English (Fourth Edition). Available separately.

b. Breul, Karl H. *Cassell's New German and English Dictionary,* revised by H. T. Betteridge. London: Cassell and Company, Ltd.; New York: Funk and Wagnalls Company; 1957. 1250 pp.

c. Mansion, J. E. *Heath's Standard French and English Dictionary.* Boston: D. C. Heath and Company, 1934-1939.[8] 2 vols. with supplement (1961).

d. Girard, Denis. *Cassell's New French Dictionary: French-English, English-French.* New York: Funk and Wagnalls Company, 1962. 1417 pp.

e. Peers, Edgar A. *Cassell's Spanish-English, English-Spanish Dictionary.* London: Cassell and Company, Ltd.; New York: Funk and Wagnalls Company; 1960. 1477 pp.

f. *Langenscheidt's Russian-English, English-Russian Dictionary.* London: Methuen and Company, 1964. 505 pp.

Technical foreign dictionaries.

Depending on typography and comprehensiveness, such dictionaries vary widely from pocket size to several large volumes. Examples of bi-lingual dictionaries are:

a. De Vries, Louis, and Clason, W. E. *Dictionary of Pure and Applied Physics.* New York: American Elsevier Publishing Company, 1963-1964. 2 vols. Vol. 1: German-English; Vol. 2: English-German.

b. De Vries, Louis. *Technical and Engineering Dictionary.* New York: McGraw-Hill Book Company. 3 vols., as follows:
Vol. 1: German-English, 1965. (Second Edition); Vol. 2: English-German, 1954; Supplement, 1959.

c. Hyman, Charles J., and Idlin, R., editors. *Dictionary of Physics and Allied Sciences.* London: Peter Owen, 1958-1962. 2 vols. (by respective editors). Vol. 1: German-English; Vol. 2: English-German.

d. Hyman, Charles J. *German-English and English-German Electronics Dictionary.* New York: Consultants Bureau, 1965. 182 pp.

e. Bindman, Werner. *Dictionary of Semiconductor Physics and Electronics; English-German; German-English.* Oxford, New York, etc.: Pergamon Press, 1966. 615 pp.

f. De Vries, Louis. *French-English Science Dictionary for Students*

8. Reprinted 1947-1948 with corrections.

in Agricultural, Biological and Physical Sciences. (Third Edition.) New York: McGraw-Hill Book Company, 1962. 655 pp.

g. Guinle, R. L. *A Modern Spanish-English and English-Spanish Technical and Engineering Dictionary.* New York: E. P. Dutton and Company, 1938. 311 pp.

h. Emin, Irving. *Russian-English Physics Dictionary.* New York: John Wiley and Sons, 1963. 554 pp.

i. Karpovich, E. A. *Russian-English Atomic Dictionary.* (Second Edition.) New York: Technical Dictionaries Company, 1959. 317 pp.

j. Zimmerman, M. G. *Russian-English Translators Dictionary.* New York: Plenum Press, 1967. 294 pp.

Larger format bi-lingual dictionaries often useful because of their comprehensiveness are:

a. Webel, A. *A German-English Dictionary of Technical, Scientific and General Terms.* (Third Edition.) London: Routledge and Kegan Paul, 1952. 939 pp.

b. Kettridge, Julius O. *French-English and English-French Dictionary of Technical Terms and Phrases.* (Third Edition.) London: Routledge and Kegan Paul, 1948. 2 vols.

Multi-lingual lexicons enable one to work among several languages simultaneously.

An old stand-by is:

Hoyer-Kreuter Technologisches Wörterbuch, herausgegeben von Alfred Schlomann. (Sechste Auflage.) Berlin: Springer, 1932. 3 vols. Vol. 1: Deutsch-English-Französisch; Vol. 2: English-German-French; Vol. 3: Francais-Allemand-Anglais.

More recent are these selections from the extensive series of Elsevier multilingual dictionaries, furnishing French, Spanish, Italian, Dutch and German equivalents of English/American terms:

a. Clason, W. E. *Dictionary of General Physics.* Amsterdam, New York, etc.: Elsevier Publishing Company, 1962. 859 pp.

b. Clason, W. E. *Dictionary of Nuclear Science and Technology.* Amsterdam, New York, etc.: Elsevier Publishing Company, 1958. 914 pp.

A Russian supplement is available.

c. Clason, W. E. *Dictionary of Electronics and Waveguides.* (Second Edition.) Amsterdam, New York, etc.: Elsevier Publishing Company, 1966. 833 pp.

Russian and Swedish supplements to the first edition are available.

d. Clason, W. E. *Supplement to the Elsevier Dictionaries of Elec-*

tronics, Nucleonics and Telecommunication. Amsterdam, New York, etc.: Elsevier Publishing Company, 1963. 633 pp.
This serves to update their dictionaries of: Television, radar and antennas; Nuclear science and technology; Amplification, modulation, reception and transmission; Electronics and waveguides (1st ed.); Cinema, sound and music; and Automation, computers, control and measuring.

Other multilingual dictionaries include:

a. Sube, R. *Dictionary of Nuclear Physics, English-German-French-Russian.* Oxford, New York, etc.: Pergamon Press, 1961. 1606 pp.

b. United Nations. Terminology Section. *Atomic Energy: Glossary of Technical Terms, in English, French, Spanish and Russian.* New York: Columbia University Press, 1958. 215 pp.

c. Neidhardt, P., editor. *Technical Dictionary of Electronics—English, French, German, Russian.* Oxford, New York, etc.: Pergamon Press, 1967. 1660 pp.

A "reading-knowledge" of a foreign scientific language may be obtained by using a standard grammar with books like the following:

German

a. De Vries, Louis. *Guide to Scientific German.* (Revised Edition.) New York: Rinehart and Company, 1953. 57 pp.
Grammar résumés are followed by illustrative passages. "In no other workbook will an instructor find such a variety of troublesome constructions presented in so few pages."

b. Eichner, Hans, and Hein, Hans. *Reading German for Scientists.* New York: John Wiley and Sons, 1959. 207 pp.
Difficult constructions are clarified in conjunction with graded readings in physics and chemistry.

c. Condoyannis, George E. *Scientific German.* New York: John Wiley and Sons, 1957. 164 pp.
Structural elements of German are described.

d. Calthrop, John E. *A German Physics Reader.* London: William Heinemann, 1943. 83 pp.
Passages on the history of physics with parallel translations opposite are followed by exercises for translation.

e. Baravalle, Hermann von. *Physik; Wärmelehre, Magnetismus, Elektrizität,* edited by S. H. Muller. Boston: D. C. Heath and Company, 1946. 50 pp.
Terms appearing in the German text are defined on the page opposite.

Russian

a. Perry, James W. *Scientific Russian.* (Second Edition.) New York: Interscience Publishers, 1961. 565 pp.

b. Emery, Mary A., and Emery, Serge A. *Scientific Russian Guide.* New York: McGraw-Hill Book Company, 1961. 191 pp.
Graded science readings from modern sources.

c. Waring, Alan G. *Russian Science Grammar.* Oxford, New York, etc.: Pergamon Press, 1967. 174 pp.

d. Holt, Arthur. *Scientific Russian: Grammar, Reading and Specially Selected Scientific Translation Exercises.* New York: John Wiley and Sons, 1962. 195 pp.

e. Starchuk, O., and Chanal, H. *Essentials of Scientific Russian.* Reading, Mass.: Addison-Wesley Publishing Company, 1963. 300 pp.
Grammar and readings are presented.

French

a. Locke, William N. *Scientific French.* New York: John Wiley and Sons, 1957. 112 pp.
Structural elements of French are described.

b. Moffatt, Christopher W. P. *Science French Course,* revised by Noel Corcoran. (Fourth Edition.) New York: Chemical Publishing Company, 1951. 332 pp.

c. Alberse, James D. *Chemical French Reader.* Boston: D. C. Heath and Company, 1940. 117 pp.
The passage on X-rays, radioactivity, etc., are of physics interest.

Summary

Roller's booklet on terminology provides a useful résumé of the peculiarities of scientific words, and a guide to their meaning and correct use. Definitions may be found in general or specialized dictionaries, including those of encyclopedic type, notably Thewlis. For translating, there is a choice between general foreign-language dictionaries and those of scientific nature. Combined use of both types is a common compromise after one has mastered language difficulties via grammar and chrestomathy.

CHAPTER IX

ADDITIONAL APPROACHES

Miscellaneous Aspects

Before proceeding with a topical treatment of physics based on subject content, a remaining group of aspects will be discussed, viz., presentational, philosophical, and practical.

1—Presentational

As previously indicated, outstanding scientists are usually master expositors whose firm grasp of their subject enables them to enlighten others. All physics teachers and students should strive for this clarity of expression, both oral and written. Furthermore, good exposition is enriched by teaching aids, such as demonstrations, audio-visual materials, etc.

Successful presentation involves the four skills indicated by:

Dean, Howard H., and Bryson, Kenneth D. *Effective Communication: A Guide to Reading, Writing, Speaking and Listening*. (Second Edition.) Englewood Cliffs, N. J.: Prentice-Hall, Inc., 1961. 560 pp.

Lectures.

Audience interest may be aroused and sustained by good speaking techniques, described in:

a. Darrow, Karl K. "How to Address the American Physical Society." *Physics Today,* 4 (No. 2): 4-8, February 1951; reprinted in *Physics Today,* 14 (No. 10): 20-23, October 1961.

b. Tucker, S. M. *Public Speaking for Technical Men.* New York: McGraw-Hill Book Company, 1939. 397 pp.

c. Casey, Robert S. *Oral Communication of Technical Information.* New York: Reinhold Publishing Corporation, 1958. 199 pp.

Slides can make or mar a presentation:

Van Pelt, J. R. "Lantern Slides and Such." *American Scientist,* 38: 450-455, July 1950.

Listening and learning are also stimulated by the visual appeal of demonstrations, such as those found in:

a. Sutton, Richard M. *Demonstration Experiments in Physics.* New York: McGraw-Hill Book Company, 1938. 545 pp.

General techniques are discussed (pp. 1-14), and further sources of experiments are listed (pp. 509-510).

b. Nokes, Malcolm C. *Demonstrations in Modern Physics.* London: William Heinemann, Ltd., 1952. 134 pp.

Practical instructions for demonstrations on conduction in gases; measurement of e/m; radioactivity; spectra; etc.

A widely used German compilation is:

Weinhold, Adolf F. *Physikalische Demonstrationen,* bearbeitet von L. Weinhold and M. Günther. (7. Auflage.) Leipzig: J. A. Barth, 1931. 740 pp.

Simple equipment [1] is featured in experiments selected from the English *School Science Review,* comprising *The Science Masters' Book.* Four series have appeared since 1931, part 1 of each covering Physics. (London: John Murray.)

Optics has its own collection:

Johnson, B. K. *Lecture Experiments in Optics.* London: Edward Arnold and Company, 1930. 112 pp.

See also C. H. Palmer's *Optics* . . .

The Center for Educational Apparatus in Physics has compiled two useful booklets:

a. *Physics Apparatus, Experiments, and Demonstrations: A Bibliographic Guide.* New York: American Institute of Physics, 1965. 81 pp. (Publication No. R-179.)

b. *Physics Experiments and Demonstrations,* selected from the *American Journal of Physics* 1933-1964: *An Annotated Subject Index.* New York: American Institute of Physics, 1965. 100 pp. (Publication No. R-182.)

The most time-honored lecture series has been held annually since 1826 at the Royal Institution of London, as described in an interesting article.[2] The lectures were designed for "a juvenile auditory" but all ages gathered to enjoy the brilliant expositions and spectacular demonstrations. Abbreviated references to some of the books based on the series follow:

Andrade, E. N. *Engines.* Bell, 1928. 267 pp.

---

1. Also employed in: K. M. Swezey, *After Dinner Science.* New York: McGraw-Hill Book Company, 1948. 182 pp.; and in I. M. Freeman, *Invitation to Experiment.* New York: E. P. Dutton and Company, 1940. 238 pp.
2. T. Martin, "Christmas Lectures at the Royal Institution." *Physics Today,* 1 (No. 3): 10-15, July 1948. (For paintings of lectures, see *American Journal of Physics,* 8: 387-390, December 1940.)

Bragg, W. H. *Concerning Nature of Things.* Bell, 1925. 232 pp.
Bragg, W. H. *Universe of Light.* Bell, 1933. 283 pp.
Bragg, W. H. *World of Sound.* Bell, 1920. 196 pp.
Bragg, W. L. *Electricity.* Bell, 1936. 286 pp.
Faraday, M. *Various Forces of Matter.* (3d ed.) Griffin, 1861. 179 pp.
Fleming, J. A. *Waves and Ripples.* (Rev. issue) Sheldon, 1923. 299 pp.
Wood, A. *Sound Waves.* Blackie, 1930. 152 pp.

Of special interest is the following one, which the Physical Science Study Committee reissued as a Science Study paperback through Doubleday:

Boys, Sir Charles V. *Soap Bubbles and the Forces Which Mold Them.* Garden City, N. Y.: Doubleday and Company, 1959. 156 pp. (This is also available through Thomas Y. Crowell Company, 1962. 280 pp.)

See also Michael Faraday's *Advice to a Lecturer* on diction, notes, demonstrations, etc., published by the Royal Institution in 1961.

Demonstrations may also be of display type, often operable by the visitor [3] and accompanied by descriptive lectures, as in:

a. Lemon, Harvey B., and Marshall, Fitz-Hugh. *The Demonstration Laboratory of Physics at the University of Chicago.* Chicago: University of Chicago Press, 1939. 127 pp.
This and similar museum projects are discussed as educational tools.

b. White, Harvey E., *et al.* "Quantitative Demonstration Exhibits and a New Low-Cost Physics Laboratory." *American Journal of Physics,* 34: 660-664, August 1966.
For self-instruction; operable by students.

Finally, as really good teachers know, even the duly scheduled lecture-class may partake of the excitement and fascination of skillful presentations:

Feynman, Richard P.; Leighton, Robert B.; and Sands, Matthew. *The Feynman Lectures on Physics.* Reading, Mass.: Addison-Wesley Publishing Company, 1964-1965. 3 vols.

Writings.

Bibliographies on various aspects of technical writing are available:

a. United Nations Educational, Scientific and Cultural Organiza-

3. See R. P. Shaw, "The Progressive Exhibit Method." *American Physics Teacher,* 7: 165-172, June 1939.

tion. *Bibliography of Publications Designed to Raise the Level of Scientific Literature.* New York: Columbia University Press, 1963. 83 pp.
Some chapters are: Technique of technical writing; Books on editing, printing and publishing; Readings in science for technical authors; and Handbooks for authors.

b. Society of Technical Writers and Publishers, Inc. New York Chapter. *Bibliography of Technical Writing, 1945-1961.* (Third Edition.) New York: The Society, 1962. 53 pp.
From various subject headings, e.g., editing, numbered reference is made to the main list.

Roller's interesting article [4] may be supplemented by:

a. Crouch, W. G., and Zetler, Robert L. *A Guide to Technical Writing.* (Second Edition.) New York: Ronald Press Company, 1954. 441 pp.

b. Gloag, John. *How to Write Technical Books.* London: George Allen and Unwin, Ltd., 1950. 159 pp.

c. Menzel, Donald H.; Jones, Howard M.; and Boyd, Lyle G. *Writing a Technical Paper.* New York: McGraw-Hill Book Company, 1961. 132 pp.

d. Morris, Jackson E. *Principles of Scientific and Technical Writing.* New York: McGraw-Hill Book Company, 1966. 257 pp.

e. Weil, B. H., editor. *Technical Editing.* New York: Reinhold Publishing Corporation, 1959. 278 pp.

f. Koefod, Paul E. *The Writing Requirements for Graduate Degrees.* Englewood Cliffs, N. J.: Prentice-Hall, Inc., 1964. 268 pp.

For various aspects of the technical report, and increasingly important medium of communication, consult:

a. Fry, Bernard M. *Library Organization and Management of Technical Reports Literature.* Washington, D. C.: Catholic University of America Press, 1953. 140 pp.

b. Weil, B. H., editor. *The Technical Report; Its Preparation, Processing, and Use in Industry and Government.* New York: Reinhold Publishing Corporation, 1954. 485 pp.

c. Waldo, Willis H. *Better Report Writing.* (Second Edition.) New York: Reinhold Publishing Corporation, 1965. 288 pp.

See also index under Reports, technical; and Houghton's *Technical Information Sources.*

When a book is being prepared with a particular publisher in mind,

---

4. D. Roller, "Technical Writing and Editing." *American Journal of Physics,* 13: 99-105, April 1945.

the author should observe that firm's accepted style, usually outlined in its own manual. Suggestions for style of articles may be found on the covers of certain journals, while an important group of physics periodicals has its own guide:

American Institute of Physics. *Style Manual for Guidance in the Preparation of Papers for Journals Published by the American Institute of Physics.* (Second Edition.) New York: The Institute, 1959. 42 pp.

Some of the sections: Preparation of a scientific paper; General style; Presentation of mathematical expressions; Preparation of illustrations; and Short history of a manuscript.

Good general manuals on style include:

a. *Words Into Type,* based on studies by Marjorie E. Skillin, Robert M. Gay, *et al.* (Revised Edition.) New York: Appleton-Century-Crofts, Inc., 1964. 596 pp.

b. Mawson, C. O. S. *Style-book for Writers and Editors.* New York: Thomas Y. Crowell Company, 1926. 213 pp.

The basis of correct style is stressed. "The style-books of the foremost publishers in the United States and Great Britain have been critically examined, in order that the most approved usages might be followed."

Other style manuals are listed, together with comments on prospective publications, by Scherr.[5] *See also* ACRL [6] Monograph No. 8 (1953): *Bibliographical Style Manuals; A Guide to their Use in Documentation and Research,* by Mary R. Kinney.

Timely assistance for anyone engaged in typing books, articles or reports in accepted form is provided by:

a. Dunford, Nelson J. *A Handbook for Technical Typists.* New York: Gordon and Breach, 1964. 136 pp.

b. Stafford, Allison R., and Culpepper, Billie J. *The Science-Engineering Secretary; A Guide to Procedure, Usage and Style.* Englewood Cliffs, N. J.: Prentice-Hall, Inc., 1963. 338 pp.

c. Hawley, Gessner G., and Hawley, Alice W. *Hawley's Technical Speller.* New York: Reinhold Publishing Corporation, 1955. 146 pp.

Spelling and syllabication for 8000 words.

d. Miles, Aetna. *Technical Speller and Definition Finder.* Indianapolis, Ind.: Howard W. Sams and Company, 1965. 283 pp.

Spelling of 50,000 terms and references to further information.

---

5. J. M. Scherr, "Authorship Without Tears; A Guide to Authors' Handbooks and Publishers' Style Manuals." *Bulletin of Bibliography,* 18: 225-227, May-August 1946.
6. Association of College and Research Libraries.

e. Shaw, Harry L. *Punctuate It Right!* New York: Barnes and Noble, Inc., 1963. 176 pp.

"A good subject index is necessary in all technical works," states a McGraw-Hill style manual, going on to quote a reviewing medium:

> The publisher and the author did not think well enough of this book to supply it with a suitable index. We feel, therefore, that it is hardly worthy of a review in our columns.

The necessary techniques may be gleaned from:

Collison, Robert L. W. *Indexing Books; A Manual of Basic Principles.* New York: John De Graff, Inc., 1962. 96 pp.

Negotiations with publishers are outlined in:

Baumol, William J., and Heim, Peggy. "On Contracting with Publishers: or What Every Author Should Know." *AAUP Bulletin,* 53: 30-46, March 1967.

Literary aspects of technical literature may be derived from:

a. Law, Frederick H., editor. *Science in Literature; A Collection of Literary Scientific Essays.* New York: Harper and Brothers, 1929. 364 pp.

Prepared for high school use, this emphasizes the spirit and style of excerpted writings, such as Madame Curie's, "The Discovery of Radium." The author states (p. xvi):

> . . . The purpose of this book is not at all the presenting of facts, or the explanation of theories, but wholly the literary power, the driving force, the uplifting inspiration, of the words of scientific workers.

b. Eastwood, Wilfred, editor. *Science and Literature; The Literary Relations of Science and Technology; An Anthology.* (First and Second Series.) London: The Macmillan Company, 1957-1960. 2 vols.

c. Eastwood, Wilfred, editor. *A Book of Science Verse: The Poetic Relations of Science and Technology: An Anthology.* London: The Macmillan Company, 1961. 279 pp.

d. Evans, Benjamin I. *Literature and Science.* London: George Allen and Unwin, Ltd., 1954. 114 pp.

This outlines historical relationships between science and literature since the Renaissance.

e. Cadden, John J., and Brostowin, Patrick R., editors. *Science and Literature—A Reader.* Boston: D. C. Heath and Company, 1964. 310 pp.

Excerpts and essays on the interrelationship.

f. Plotz, Helen, compiler. *Imagination's Other Place; Poems of*

Science and Mathematics. New York: Thomas Y. Crowell Company, 1955. 200 pp.

See also "Resource Letter SL-1 on Science and Literature," by Marjorie Nicolson. *American Journal of Physics*, 33: 175-183, March 1965.

Attractiveness of presentation does not depend upon artificial embellishment, according to Tripp, who states:

> It is the first duty of the monograph writer to estimate the value, either actual or potential, of recent work upon the subject of which he writes: he must pick out the plums to save others from the indigestion that follows eating the whole pie. Further, in addition to being accurate, his work must be presented in a form both assimilable and attractive; in other words, he must show that lucid exposition can be achieved by the use of few words, if they are rightly chosen, and that attractive presentation is attained rather by clear thinking than by superficial display.[7]

Finally, attention should be paid to the composition of the whole printed page, for "the visual task of reading is created by those who plan printed material," as expounded in:

Luckiesh, Matthew, and Moss, Frank K. *Reading as a Visual Task*. New York: D. Van Nostrand Company, 1942. 428 pp.

See also index under Reading skill.

Pictorial devices.

Various estimates have been ventured concerning the number of printed words that one good illustration equals. Scientific illustration from several viewpoints is described in:

a. Ridgway, John L. *Scientific Illustration*. Stanford University, Cal.: At the University Press, 1938. 173 pp.
This outlines "effective methods of preparation and the proper fitting, assembling, and display of illustrations designed for scientific publications."

b. Clarke, Carl D. *Illustration; Its Technique and Application to the Sciences*. Baltimore: The John D. Lucas Company, 1940. 386 pp.
Examples are drawn chiefly from chemistry and physiology.

c. Blaker, Alfred A. *Photography for Scientific Publication; A Handbook*. San Francisco: W. H. Freeman and Company, 1965. 158 pp.

---

7. E. H. Tripp in preface to J. A. Radley and J. Grant, *Fluorescence Analysis in Ultra-Violet Light*, p. vi. (Fourth Edition.) London: Chapman and Hall, Ltd., 1954.

See also books by M. Cagnet, W. Gentner, C.-N. Martin, C. F. Powell, and G. D. Rochester (via author index).

The American Institute of Physics has a Photographic Depository which supplies prints "at nominal cost to authors, publishers and organizations having a legitimate use for them in publications pertaining to physics."

Drafting in conformity with U. S. Patent Office requirements is shown by:

Radzinsky, Harry. *Making Patent Drawings.* New York: The Macmillan Company, 1945. 96 pp.

Physics graphs and diagrams constitute an unusual German compilation designed to accompany a textbook:

Auerbach, Felix. *Physik in Graphischen Darstellungen.* (Zweite Auflage.) Leipzig: B. G. Teubner, 1925. 257: 29 pp.
There are 1557 pictorial representations of physical entities, such as hysteresis loops, Van der Waals constant plots, flame pictures, etc.

Portraits of physicists are included in a superb portfolio:

Crew, Henry. *Portraits of Famous Physicists.* New York: Scripta Mathematica, Yeshiva College, 1942. 12 folders.
Brief biographical sketches accompany the portraits of Ampère, Clausius, Faraday, Fresnel, Galileo, Gibbs, Hertz, Huygens, Joule, Maxwell, Newton, Rowland.

Similar portfolios that include some physicists are as follows:

Journal of Chemical Education. Distinguished Chemists. Series A, B, C. Easton, Pa.: Mack Printing Company, 1935-1936. 3 folders.

Biographical Memoirs (National Academy of Sciences), and *Biographical Memoirs of Fellows of the Royal Society of London,* both feature portraits.

Other sources of portraits are the general and biographical encyclopedias, histories, biographies, and the following index:

A. L. A. Portrait Index, edited by W. C. Lane and N. E. Browne. Washington, D. C.: Library of Congress, 1906. 1600 pp.

N. O. Ireland's *Index to Scientists* also indicates location of portraits.

See also the Nobel laureates' portraits in Weber, Manning and White's College Physics.

"Reproduction of Prints, Drawings and Paintings of Interest in the History of Physics" have been appearing in the *American Journal of Physics* and its predecessor (*American Physics Teacher*) from June 1938 to September 1956, except 1950-1951. They may be located under name of contributor, E. C. Watson, in the annual author indexes.

Schaeffer [8] finds similar historic interest in certain postage-stamp designs.

Audio-visual materials, such as motion-pictures, slides, filmstrips, recordings, etc., are increasingly gaining prominence as educational tools.

Earlier lists of these non-book materials (*Educational Film Guide, Filmstrip Guide,* etc.) have been superseded by:

Educational Media Index, a project of the Educational Media Council. New York: McGraw-Hill Book Company, 1964. 14 vols.
Of immediate interest is Vol. 11: Science and engineering. Symbols indicate form of material.

For 16 mm. films in particular see:

National Information Center for Educational Media. *Index to 16 mm. Educational Films.* New York: McGraw-Hill Book Company, 1967. 955 pp.

As a separate supplement to its *National Union Catalog,* the Library of Congress still publishes *Motion Pictures and Filmstrips* quarterly, with annual and quinquennial cumulations.

Other sources of physics films include:

a. Weber, Robert L. "Films for Students of Physics." *American Journal of Physics,* 29: 222-233, April 1961; also Supplement 1, *ibid.,* 30: 321-327, May 1962.
A very useful list of films supplementing physics instruction at the college level.

b. *Educators Guide to Free Science Materials.* (Seventh Edition.) Randolph, Wisc.: Educators Progress Service, 1966. 358 pp.
Includes a section of physics films, as well as other science films, filmstrips, exhibits, etc.

Various engineering firms, e.g., General Electric Company, produce and distribute free films incorporating maximum information and minimum advertising.

Studies have been made of films as instructional tools, including:

a. Gale, Grant O. *A Study of the Teaching of Physics by Film and Television.* New York: American Institute of Physics, 1958. 125 pp.
The American Association of Physics Teachers and the American

8. H. F. Schaeffer, "Philately for Physicists." *American Physics Teacher,* 6: 21-24, February 1938.

Institute of Physics sponsored this evaluation program (1957-1958).

b. Tendam, D. J., *et al.* "Production of Instructional Films with University Facilities." *American Journal of Physics,* 30: 517-521, July 1962; and "An Experimental Evaluation of the Use of Instructional Films in College Physics." *Ibid.,* 30: 594-601, August 1962. Twenty sound motion-pictures were made of demonstration experiments in mechanics and wave motion. They proved to be satisfactory instructional media.

A more extensive physics series at the high school level was filmed live as televised:

Introductory Physics on Film, taught by Harvey E. White. Wilmette, Ill.: Encyclopaedia Britannica Films, Inc., 1957. 162 half-hour lessons; black and white, or color.
These films were reviewed in *American Journal of Physics,* 26: 508-512, October 1958; and *ibid,* 26: 647-650, December 1958.

Educational Services, Inc., of Watertown, Mass., has produced many films for the Physical Science Study Committee's high-school course.

Films, as well as books, are reviewed currently in the *American Journal of Physics* and *The Physics Teacher.*

2—Philosophical

Although there is a borderland between physics and philosophy, it is not as wide and uncharted as the casual observer might suppose. The following statements indicate narrowing of the gap:

> In a still broader sense physics, in the hands of several distinguished scholars, is showing a tendency to justify its original title of "natural philosophy." Certainly it is true, as it has been in the past, that the discoveries of science, however abstract they may be, have a profound bearing upon our thought. More and more do the findings of modern physics, as for example relativity and atomic theory, bring to philosophy contributions whose importance can hardly be over-estimated.[9]

"Physics and philosophy thus cooperate in giving us a complete knowledge of the universe."[10]

Philosophy of science.

A thorough understanding of physics methodology and concepts should precede attempts at philosophical reasoning. In addition to

---

9. A. B. Crawford and S. H. Clement, *The Choice of an Occupation,* p. 159. New Haven, Conn.: Yale University Press, 1932.
10. O. A. Grosselin, "The Relation of Physics to Philosophy." *American Journal of Physics,* 9: 285-290, October 1941.

several books previously cited at the beginning of Chapter V, see the following:

a. Conant, James B. *Science and Common Sense.* New Haven, Conn.: Yale University Press, 1951. 271 pp.
The author interprets science for the layman, less pedagogically than in his earlier *On Understanding Science.*

b. Watson, W. H. *Understanding Physics Today.* Cambridge, England: At the University Press, 1963. 219 pp.

c. Weizsäcker, Carl F. von, and Juilfs, Johs. *Contemporary Physics.* New York: George Braziller, Inc., 1962. 150 pp.
This is a revision of their *Rise of Modern Physics.*

d. Bridgman, Percy W. *The Logic of Modern Physics.* New York: The Macmillan Company, 1927. 228 pp.
The author states (p. x):

> It is the attempt of this essay to give a more or less inclusive critique of all physics. Our problem is the double one of understanding what we are trying to do and what our ideals should be in physics, and of understanding the nature of the structure of physics as it now exists. These two ends are together furthered by an analysis of the fundamental concepts of physics; an understanding of the concepts we now have discloses the present structure of physics, and a realization of what the concepts should be involves the ideals of physics.

e. Bridgman, Percy W. *The Nature of Physical Theory.* Princeton, N. J.: Princeton University Press, 1936. 138 pp.

See also "Resource Letter PhM-1 on Philosophical Foundations of Classical Mechanics," by Mary Hesse. *American Journal of Physics,* 32: 905-911, December 1964.

A few of the many philosophical treatments may be cited:

a. Planck, Max. *The Philosophy of Physics.* New York: W. W. Norton and Company, 1936. 128 pp.
Contents are: Physics and world philosophy; Causality in nature; Scientific ideas: their origin and effects; Science and faith.

b. Jeans, Sir James H. *Physics and Philosophy.* Cambridge, England: At the University Press, 1943. 222 pp.

c. Frank, Philipp. *Philosophy of Science.* Englewood Cliffs, N. J.: Prentice-Hall, Inc., 1957. 394 pp.

d. Eddington, Sir Arthur S. *The Philosophy of Physical Science.* Cambridge, England: At the University Press, 1949. 230 pp.

e. Čapek, Milič. *The Philosophical Impact of Contemporary Physics.* Princeton, N. J.: D. Van Nostrand Company, 1961. 414 pp.

f. Bunge, Mario, editor. *Studies in Foundations, Methodology and*

Philosophy of Science. New York: Springer-Verlag, 1967- 4 vols., in process.

Interesting collections of readings are available:

a. Wiener, Philip P. *Readings in Philosophy of Science; Introduction to the Foundations and Cultural Aspects of the Sciences.* New York: Charles Scribner's Sons, 1953. 645 pp.

b. Feigl, Herbert, and Brodbeck, May, editors. *Readings in the Philosophy of Science.* New York: Appleton-Century-Crofts, Inc., 1953. 811 pp.
A comprehensive subject-bibliography is included.

c. Kahl, Russell, editor. *Studies in Explanation; A Reader in the Philosophy of Science.* Englewood Cliffs, N. J.: Prentice-Hall, Inc., 1963. 363 pp.

d. Danto, Arthur, and Morgenbesser, Sidney, editors. *Philosophy of Science; Readings.* New York: Meridian Books, 1960. 477 pp.

Psychological aspects are reflected in:

Stevens, Blamey. *The Psychology of Physics.* Manchester, England: Sherratt and Hughes, 1939. 282 pp.
The author develops his perceptual theory, maintaining that the laws of physics are subjective rather than objective. "Perceptual theory thus supplies us with a fundamental system to which all phenomena may be referred. Without such perceptual basis physics is a collection of detached laws."

Social implications of science.

Selected books that reveal appreciation of science's contribution to civilization, and, more recently, apprehension concerning its destructive potentialities, follow:

a. Lindsay, Robert B. *The Role of Science in Civilization.* New York: Harper and Row, 1963. 318 pp.
Portrays the scientist as a creator and observer.

b. Mumford, Lewis. *Technics and Civilization.* New York: Harcourt, Brace and Company, 1934. 495 pp.

c. Cohen, I. Bernard. *Science Servant of Man.* Boston: Little, Brown and Company, 1948. 362 pp.

d. Bridgman, Percy W. *Reflections*[11] *of a Physicist.* New York: Philosophical Library, 1950. 392 pp.

---

11. Concerning science and society, as well as operationalism: "the meaning of any concept is to be sought in the operations, whether physical or mental, which are performed in making application of that concept."

e. Jaki, Stanley L. *The Relevance of Physics.* Chicago: University of Chicago Press, 1966. 604 pp.
Develops "awareness that physics molds history and culture not only by its discoveries but also by the state of mind it fosters."

f. Holton, Gerald, editor. *Science and Culture: A Study of Cohesive and Disjunctive Forces.* Boston: Houghton Mifflin Company, 1965. 348 pp.

g. Woolf, Harry, editor. *Science as a Cultural Force.* Baltimore: The Johns Hopkins Press, 1964. 110 pp.

See also the quarterly UNESCO journal, *Impact of Science on Society,* and the U. N. *International Bibliography of Atomic Engery,* Vol. 1.

3—Practical

Practical aspects of physics include its applications in industry, its role in the home, and its choice as a career by the embryo scientist.

Industrial applications.

That "Physics is the foundation upon which most lines of engineering are built" [12] is amply demonstrated by the following:

a. *Physics in Industry.* New York: American Institute of Physics, 1937. 290 pp.
This compilation of papers by K. T. Compton, E. C. Sullivan, Zay Jeffries, *et al.,* marked the Institute's fifth anniversary. Topics connected with the glass, metal, petroleum, building, electrical and other industries were treated, followed by a discussion of the training of physicists from educator's and employer's viewpoints. Compton points out (p. x):

> Not all who are physicists call themselves by that name and therein lies one reason why the average citizen has little comprehension of what physics really is. There has been a very interesting historic trend in physics by which great branches of its specialized interests have been appropriated by special groups of physicists who call themselves engineers just as soon as a systematic method of applying physical principles to advantageous ends has been developed in a specialized field.

b. "The Role and Training of the Physicist in Industry." *Physics Today,* 13 (No. 1): 23-56, January 1960.
Papers presented at an American Institute of Physics symposium in

---

12. J. G. McGuire and H. W. Barlow, *An Introduction to the Engineering Profession,* p. 165. (Second Edition.) Cambridge, Mass.: Addison-Wesley Press, 1951.

1959, covering the relation of physics to the missile, satellite and communication fields; the petroleum, automotive and optical industries; and high polymer and other materials.

c. Institute of Physics and the Physical Society. *Physics in Industry.* London: The Institute . . . , 1923 to date.

These publications appear irregularly, and include lectures between 1923 and 1933, later symposia, and now monographs on applied physics. One entitled "Magnetism" (1938) has an interesting comment by W. L. Bragg (p. vi):

> One cannot guarantee that the pure scientist, when made acquainted with the problems with which industry is faced, will immediately have a "brain wave" which will enable him to suggest a solution. One can say, however that unless he has some knowledge of these problems the chances of his being of assistance are very remote. By helping each other to grasp the main essential problems, we create the conditions in which inspiration is most likely to come.

What physics has been doing for industry and the fine arts is also interestingly recounted in:

Richardson, Edward G. *Physical Science in Art and Industry.* (Second Edition.) London: The English Universities Press, 1946. 299 pp.

More specialized surveys are:

a. Bosworth, R. C. L. *Physics in Chemical Industry.* London: The Macmillan Company, 1950. 928 pp.

b. Starling, Sydney G. *Electricity in the Service of Man,* revised by H. J. Gray. (Second Edition.) New York: Longmans, Green and Company, 1949. 255 pp.

c. *Applied Physics: Electronics; Optics; Metallurgy.* Boston: Little, Brown and Company, 1948. 456 pp.

World War II defense linked with scientific potential.

Industrial management relationships are examined in:

a. Kornhauser, William. *Scientists in Industry; Conflict and Accommodation.* Berkeley and Los Angeles: University of California Press, 1962. 230 pp.

Discusses professional goals, controls, organizations, incentives and influence, with respect to strains and adaptations.

b. Kuhn, James W. *Scientific and Managerial Manpower in Nuclear Industry.* New York: Columbia University Press, 1966. 209 pp.

Effects of manpower policies upon technical change.

c. Princeton University. Industrial Relations Section. *The Sci-*

entist in American Industry; Some Organizational Determinants in Manpower Utilization, by Simon Marcson. New York: Harper and Row, 1960. 158 pp.

See also index under Manpower, technical; and Industrial physics.

Atomic energy developments have made it clear that "No engineer can now say with any assurance what aspects, if any, of physical or chemical research lie outside his sphere. Never before has it been truer than now that the physics of to-day is the engineering of to-morrow." [13]

In the other direction as well, backward through time, physics helps unearth the relics of antiquity:

Aitken, M. J. *Physics and Archaeology.* New York: Interscience Publishers, 1961. 194 pp.
Intriguing techniques include electromagnetic and resistivity surveying; magnetic, thermoluminescent and radiocarbon dating; and use of proton and differential magnetometers, the beta-ray back-scatter meter, and emission and X-ray fluorescent spectrometers.

Physics even reaches out to trap the criminal, as described in:

O'Hara, Charles E., and Osterburg, James W. *An Introduction to Criminalistics; The Application of the Physical Sciences to the Detection of Crime.* New York: The Macmillan Company, 1949. 705 pp.
Of particular physics interest are discussions of the use of ultraviolet, infrared, and X-rays; photomicrography; measurement of distances, speed or force in the examination of skid marks, shattered glass, etc.

See also the final sections of the Topical Approach discussions, e.g., Applied Mechanics, Applied Thermodynamics, etc.; and index under Research, experimental.

Household applications.

The following textbooks, addressed primarily to home economics students, link physics principles with everyday experiences:

a. Whitman, Walter G. *Household Physics.* (Third Edition.) New York: John Wiley and Sons, 1939. 436 pp.
The author asserts (p. v):

> So rapid is progress in practical science that the average home of today has much equipment that was either unknown or was in the early stages of development ten years ago. Every year man in a progressive community becomes more and more dependent on devices of his own creation which make use of the principles of physics. . . . There is

13. T. W. Chalmers, *Historic Researches,* p. 2. New York: Charles Scribner's Sons, 1952.

no question but that greater knowledge of the underlying principles and an understanding of the operation of many of these household devices will increase their efficiency and cut down the repair bills. It is also a matter of satisfaction to many to understand "how it works."

b. Avery, Madalyn. *Household Physics.* (Third Edition.) New York: The Macmillan Company, 1955. 472 pp.

c. Osborn, Frederick A. *Physics of the Home.* (Third Edition.) New York: McGraw-Hill Book Company, 1935. 441 pp.

d. Peet, Louise J., and Thye, Lenore S. *Household Equipment.* (Fifth Edition.) New York: John Wiley and Sons, 1961. 357 pp.

Physics as a vocation.

Physics, today and tomorrow, is appraised in:

National Research Council, Physics Survey Committee. *Physics: Survey and Outlook;* a report on the present state of United States physics and its requirements for future growth. *Supplement* on the subfields of physics. Washington, D. C.: The Council, 1966. 119: 165 pp. (NRC Publications Nos. 1295, 1295A.)
George E. Pake headed the committee which compiled this report, which is summarized in *Physics Today,* 19 (No. 4): 23-36, April 1966.

For those interested in physics as a vocation, the question "What is Physics?" is answered on pp. 1-4 of the preceding report, and more fully in the supplementary descriptions of various subfields, such as atomic, nuclear, plasma, solid-state, etc.

Further questions:

"Who Are Physicists? What Do They Do?" *Physics Today,* 19 (No. 1): 70-76, January 1966.
National Register statistics on salaries, fields, etc., summarized by Sylvia Barisch of the American Institute of Physics unit.

"What is a Physicist?" *Review of Scientific Instruments,* 15: 54-56, February 1944.
A National Roster occupational outline of the profession of physics. (See also more recent specialties lists for use with National Register of Scientific and Technical Personnel periodic questionnaires.)

Factors most important in interesting undergraduates in science as a vocation were investigated in:

Knapp, Robert H., and Goodrich, H. B. *Origins of American Scientists.* Chicago: University of Chicago Press, 1952. 450 pp.

Whether he is adapted to practice physics as a profession, and how he may obtain a position in the field, are two extremely practical problems confronting the student contemplating a career. Both aspects are delineated in:

a. Smith, Alpheus W., and Hole, Winston L. *Careers in Physics.* (Revised Edition.) Columbus, Ohio: Long's College Book Company, 1960. 310 pp.

Comprehensive coverage of all phases.

b. Pollack, Philip. *Careers and Opportunities in Physics.* New York: E. P. Dutton and Company, 1961. 159 pp.

c. Nourse, Alan E. *So You Want to be a Physicist.* New York: Harper and Row, 1963. 182 pp.

d. Clarke, Norman. *Physics as a Career.* (Second Edition.) London: Chapman and Hall, Ltd., 1959. 79 pp. (Published for the Institute of Physics.)

Briefer résumés appear in:

a. American Institute of Physics. *Physics as a Career.* New York: The Institute, 1952. 17 pp.

b. Institute for Research. *Career as a Physicist.* Chicago: The Institute, 1946. 20 pp. (Its *Careers* series, Research No. 143.)

c. Crawford, Albert B., and Clement, Stuart H. *The Choice of an Occupation,* pp. 156-160. New Haven, Conn.: Yale University Press, 1932.

d. Stewart, George W. "Physics as a Career." *Science,* n.s. 58: 275-278, October 12, 1923.

This has also been reprinted by the National Research Council as one of the "career bulletins" comprising its collection, "Opportunities for a Career in Scientific Research," 1927.

Manuals for special fields include:

a. Hartzell, Karl D. *Opportunities in Atomic Energy.* New York: Grosset and Dunlap, 1951. 143 pp.

After sketching the atomic energy program, the author discusses personnel policies, training, qualifications, salaries, types of work, etc.

b. Kamen, Ira, and Dorf, Richard H. *TV and Electronics as a Career.* New York: John F. Rider Publisher, Inc., 1951. 326 pp.

Engineering and commercial phases are emphasized.

c. Adams, Carsbie C., *et al.* *Careers in Astronautics and Rocketry.* New York: McGraw-Hill Book Company, 1962. 252 pp.

Professional placement announcements appear among the pages of various physics periodicals, notably *Physics Today, Journal of Applied Physics,* etc. Activities of the Placement Service of the American Institute of Physics include conducting an employment register at meetings to facilitate contacts.

The preparation, availability and use of scientific personnel are surveyed in:

a. *Physics Manpower 1966: Education and Employment Statistics.* New York: American Institute of Physics, 1966. 114 pp.
This graphic report, prepared by Susanne Ellis, has three sections: 1, Physics education in the U. S.; 2, Employment of physicists in the U. S.; and 3, Foreign sources of physics manpower in the U. S.

b. National Science Foundation. *Scientific and Technical Manpower Resources; Summary Information on Employment, Characteristics, Supply, and Training.* Washington, D. C.: Government Printing Office, 1965. 184 pp.

c. *Toward Better Utilization of Scientific and Engineering Talent—A Program for Action.* Washington, D. C.: National Research Council, 1964. 153 pp. (Publication No. 1191.)
James R. Killian, Jr., headed the committee which conducted the study at the President's request.

d. Payne, George L. *Britain's Scientific and Technological Manpower.* Stanford, Cal.: Stanford University Press; London: Oxford University Press, 1960. 466 pp.

See also index under Manpower, technical.

Sidelights on scientific work for official agencies are given in:

a. *Attitudes of Scientists and Engineers about Their Government Employment.* Vol. 1. Syracuse, N. Y.: Syracuse University, 1950. 223 pp.
This was compiled by the Maxwell Graduate School of Citizenship and Public Affairs for the Office of Naval Research.

b. Price, Don K. *Government and Science; Their Dynamic Relation in American Democracy.* New York: New York University Press, 1954. 203 pp.

c. Dupree, A. Hunter. *Science in the Federal Government; A History of Policies and Activities to 1940.* Cambridge, Mass.: The Belknap Press of Harvard University, 1957. 460 pp.

Scholarships and graduate fellowships are listed in:

a. *Fellowships in the Arts and Sciences,* prepared in cooperation with the Council of Graduate Schools in the United States; edited by Robert Quick. Washington, D. C.: American Council on Education, 1957- The tenth edition covers 1967-1968; 93 pp.

b. Feingold, S. Norman. *Scholarships, Fellowships and Loans.* Cambridge, Mass.: Bellman Publishing Company, 1949-1962. 4 vols. Updated by its quarterly news service.

c. *Study Abroad: International Handbook; Fellowships, Scholar-*

ships, and Educational Exchange. Paris: UNESCO, 1948- Now biennially.

See also index under Study.

The National Science Foundation awards research grants and fellowships, as reviewed in:

McMillen, J. Howard. "A Decennial Look at NSF's Research Grants Program in Physics." *Physics Today,* 13 (No. 11): 40-42. November 1960.

See also the National Science Foundation's annual summary of *Federal Funds for Research, Development and Other Scientific Activities,* and Waterman's article.[14]

For information about grant opportunities from all sources (foundations, governmental agencies, societies, organizations, etc.) see *Grant Data Quarterly,* published by Academic Media, Inc., of Los Angeles, Cal.

---

14. A. T. Waterman, "The National Science Foundation—Its Organization and Purposes." *American Journal of Physics,* 20: 73-77, February 1952.

CHAPTER X

TOPICAL APPROACH

1—General

General textbooks.

Examination of the numerous textbooks of elementary physics on library shelves reveals essential similarity of content. Format has improved greatly during recent years, making physics study more fascinating. Each author, of course, believes his book excels in arrangement, subject content, and clarity of expression. In reality, many books serve the purposes of other teachers [1] almost equally well, leading to widespread adoption and economy of effort. As it is conceded:

> Of the writing of textbooks of physics there seems to be no end. This will continue as long as science develops—as long as the detailed requirements of a scientific training become more and more extended—and as long as there are teachers who are interested in the efficient pursuit of their vocation.[2]

A representative group of general textbooks follows:

a. Resnick, Robert, and Halliday, David. *Physics.* New York: John Wiley and Sons, 1966. 1324 pp.
This is a new version of Halliday and Resnick's 1962 *Physics . . .*, of which Part II was already in second edition stage. (Parts are available separately.)

b. Rusk, Rogers D. *Introduction to College Physics.* (Second Edition.) New York: Appleton-Century-Crofts, Inc., 1960. 944 pp.

c. Sears, Francis W., and Zemansky, Mark W. *College Physics.*[3] (Third Edition.) Reading, Mass.: Addison-Wesley Publishing Company, 1960. 1024 pp.

---

1. *Cf.* J. W. McGrath, "Instructor Opinion on Characteristics of a Good General Physics Textbook." *American Journal of Physics,* 13: 309-314, October 1945. Also P. G. Roll, "Introductory Physics Textbooks." *Physics Today,* 21 (No. 1): 63-71, January 1968.
2. M. Ference, Jr.; H. B. Lemon; and R. J. Stephenson, *Analytical Experimental Physics,* p. v. (Third Edition.) Chicago: University of Chicago Press, 1956.
3. To be distinguished from their *University Physics* which introduces calculus, as does the three-volume set *Principles of Physics* by Sears alone.

d. Semat, Henry. *Fundamentals of Physics.* (Fourth Edition.) New York: Holt, Rinehart and Winston, 1966. 753 pp.

e. Smith, Alpheus W., and Cooper, John N. *The Elements of Physics.* (Seventh Edition.) New York: McGraw-Hill Book Company, 1964. 717 pp.

f. Weber, Robert L.; Manning, Kenneth V.; and White, Marsh W. *College Physics.* (Fourth Edition.) New York: McGraw-Hill Book Company, 1965. 710 pp.

Nobel laureates' portraits have long been featured.

g. Harnwell, Gaylord P., and Legge, George J. F. *Physics: Matter, Energy and the Universe.* New York: Reinhold Publishing Corporation, 1967. 577 pp.

Designed for a full-year introductory course for non-physics majors.

Many books reveal definite slants with respect to intended use or author's outlook. For example, historical considerations are embodied in:

a. Taylor, Lloyd W. *Physics; The Pioneer Science.* Boston: Houghton Mifflin Company, 1941. 847: 44 pp.

b. Priestley, Herbert. *Introductory Physics; An Historical Approach.* Boston, Mass.: Allyn & Bacon, Inc., 1958. 515 pp.

c. Holton, Gerald, and Roller, Duane H. D. *Foundations of Modern Physical Science,*[4] edited by Duane Roller. Reading, Mass.: Addison-Wesley Publishing Company, 1958. 782 pp.

With scientific subject-content paramount throughout, the drama of science is unfolded from historical, philosophical, and logical viewpoints.

For engineering students, we have:

a. Oldenberg, Otto, and Rasmussen, Norman C. *Modern Physics for Engineers.* New York: McGraw-Hill Book Company, 1966. 477 pp.

b. Weber, Robert L.; White, Marsh W.; and Manning, Kenneth V. *Physics for Science and Engineering.* (Revised Edition.) New York: McGraw-Hill Book Company, 1957. 830 pp.

For teachers:

Weber, Robert L. *Physics for Teachers; A Modern Review.* New York: McGraw-Hill Book Company, 1964. 314 pp.

Topics of current interest, centered upon the atom and space, with teaching suggestions and materials.

---

4. See also Holton's *Introduction to Concepts and Theories in Physical Science,* 1952. 650 pp.

Adapted to science survey courses are:

a. Krauskopf, Konrad B., and Beiser, Arthur. *Fundamentals of Physical Science.* (Fifth Edition.) New York: McGraw-Hill Book Company, 1966. 720 pp.

b. Gray, Dwight E., and Coutts, John W. *Man and His Physical World.* (Third Edition.) Princeton, N. J.: D. Van Nostrand Company, 1958. 672 pp.

Unifying treatments of physical properties include:

a. Newman, F. H., and Searle, V. H. L. *The General Properties of Matter.* (Fourth Edition.) New York: Longmans, Green and Company, 1948. 431 pp.

b. Champion, Frank C., and Davy, N. *Properties of Matter.* (Third Edition.) London: Blackie and Son, 1959. 334 pp.

Some books are produced under special auspices:

a. Physical Science Study Committee. *Physics.* (Second Edition.) Boston: D. C. Heath and Company, 1965. 686 pp.
The new high school course, further described under Recent Period in the prefatory overview of the Educational Approach.

b. *Berkeley Physics Course.* New York: McGraw-Hill Book Company, 1965-1967. 5 vols.
This course was planned by an interuniversity committee at the University of California, Berkeley.

Finally, for modern as opposed to classical physics there are:

a. Sproull, Robert L. *Modern Physics.* (Second Edition.) New York: John Wiley and Sons, 1963. 630 pp.

b. Leighton, Robert B. *Principles of Modern Physics.* New York: McGraw-Hill Book Company, 1959. 795 pp.

c. Richtmyer, Floyd K.; Kennard, E. H.; and Lauritsen, T. *Introduction to Modern Physics.* (Fifth Edition.) New York: McGraw-Hill Book Company, 1955. 666 pp.

d. White, Harvey E. *Modern College Physics.* (Fifth Edition.) Princeton, N. J.: D. Van Nostrand Company, 1966. 784 pp.

For theoretical physics textbooks, see Chapter VI; for atomic, Chapter X, Section 8.

Comprehensive sets.

The current systematic treatise in a long succession of "Handbuch" compilations is:

Handbuch der Physik; Encyclopedia of Physics, herausgegeben von S. Flügge. Berlin: Springer, 1955- 54 vols., in process.

Contributions are in German, English or French. This is a truly monumental work covering the entire field of physics.

Predecessors, of historic interest, were:

a. *Handbuch der Physik,* herausgegeben von H. Geiger und K. Scheel. (Vols. XXII-XXIV: Second Edition.) Berlin: Springer, 1926-1933. 24 vols. in 27, plus separate Sachregister (1929; 26 pp.)

b. *Handbuch der Experimentalphysik,* herausgegeben von W. Wien und F. Harms. Leipzig: Akademische Verlagsgesellschaft, 1926-1937. 26 vols. in 44, plus two supplementary volumes.

Several specialized handbooks mentioned under subject [5] were based on the earlier:

Winkelmann, A. *Handbuch der Physik.* (Zweite Auflage.) Leipzig: J. A. Barth, 1905-1909. 6 vols. in 7.

Another large German encyclopedia:

Handwörterbuch der Naturwissenschaften, herausgegeben von R. Dittler, G. Joos, *et al.* (Zweite Auflage.) Jena: Gustav Fischer, 1931-1935. 10 vols. plus index vol.
The articles are extensive, with many biographies and bibliographies included.

Also encyclopedic in scope:

Müller-Pouillets Lehrbuch der Physik. (11. Auflage.) Braunschweig: Friedrich Vieweg und Sohn, 1925-1934. 5 vols. in 14.

Finally, an encyclopedic work of interest to the physicist, even though portions date back to the turn of the century:

Encyklopädie der Mathematischen Wissenschaften, mit Einschluss Ihrer Anwendungen. Leipzig: B. G. Teubner, 1898-1935. 6 vols. in 23. (Vol. 4: *Mechanik,* redigiert von Felix Klein und C. H. Müller; 1901-1914, with index vol., 1935. Vol. 5: *Physik,* redigiert von A. Sommerfeld, 1903-1926.)
Among its noted contributors are Abraham, Born, Lorentz, Love, *et al.* Long bibliographies are given. A second edition is in progress (Leipzig: B. G. Teubner, 1939-). There is an incomplete French edition under the title: *Encyclopédie des Sciences Mathématiques Pures et Appliquées.*

See also the encyclopedic dictionaries of Thewlis and Glazebrook, and the McGraw-Hill, Van Nostrand, and Harper scientific encyclopedias, previously discussed.

---

5. Mechanics (Auerbach); Optics (Gehrcke); Electricity and Magnetism (Graetz).

2—Mechanics

General.

For the field of mechanics there exists a comprehensive encyclopedia of historical interest:

Auerbach, Felix, and Hort, W. *Handbuch der Physikalischen und Technischen Mechanik.* Leipzig: J. A. Barth, 1927-1931. 7 vols. in 8. This is one of several subject handbooks based on Winkelmann's *Handbuch der Physik.*

For interesting accounts of pioneer work in mechanics, see:

a. Hart, Ivor B. *The Mechanical Investigations of Leonardo da Vinci.* (Second Edition.) Berkeley, Cal.: University of California Press, 1963. 240 pp.
In superb format, Da Vinci's wide range of invention is set forth against comtemporary backgrounds.

b. Galilei, Galileo. *Dialogues Concerning Two New Sciences,* translated by Henry Crew and Alfonso de Salvio. Evanston and Chicago, Ill.: Northwestern University, 1939. 300 pp.
The translators state (p. v):

> For more than a century English speaking students have been placed in the anomalous position of hearing Galileo constantly referred to as the founder of modern physical science without having any chance to read, in their own language, what Galileo himself has to say.

The "two new sciences" are mechanics and local motions.

Energy is viewed as a central concept of physics as well as mechanics in:

a. Chalmers, Bruce. *Energy.* New York: Academic Press, 1963. 289 pp.
Some of the sections: Force; Momentum; Kinetic energy; Conversion of heat into work; Radiation; and Sources and uses of energy.

b. Theobald, David W. *The Concept of Energy.* London: E. and F. N. Spon, 1966. 192 pp.

See also index under Energy.

Of interest historically and as lucid presentations of mechanics are the various books by Edward J. Routh that were being widely used at the turn of the century. They covered the two branches (statics and dynamics), elementary and advanced, for particles and rigid bodies.

The next group of classics would include:

a. Jeans, Sir James H. *An Elementary Treatise on Theoretical Mechanics.* Boston: Ginn and Company, 1907. 364 pp.
A clear understanding of physical principles is provided.

b. Love, Augustus E. H. *Theoretical Mechanics: An Introductory Treatise on the Principles of Dynamics.* (Third Edition.) Cambridge, England: At the University Press, 1921. 310 pp.

c. Lamb, Sir Horace. *Higher Mechanics.* (Second Edition.) Cambridge, England: At the University Press, 1929. 292 pp.

d. Lamb, Sir Horace. *Statics, including Hydrostatics and the Elements of the Theory of Elasticity.* (Third Edition.) Cambridge, England: At the Univeristy Press, 1928. 357 pp.

e. Lamb, Sir Horace. *Dynamics.* (Second Edition.) Cambridge, England: At the University Press, 1928. 357 pp.

f. Whittaker, Sir Edmund T. *A Treatise on the Analytical Dynamics of Particles and Rigid Bodies.* (Fourth Edition.) Cambridge, England: At the University Press, 1937. 456 pp.

g. Ramsey, Arthur S. *Statics.* Cambridge, England: At the University Press, 1934. 296 pp.

h. Ramsey, Arthur S. *Dynamics.* (Second Edition.) Cambridge, England: At the University Press, 1951. 2 vols.

Introductory books, selected from among many:

a. Lindsay, Robert B. *Physical Mechanics.* (Third Edition.) Princeton, N. J.: D. Van Nostrand Company, 1961. 471 pp.

b. Fowles, Grant R. *Analytical Mechanics.* New York: Holt, Rinehart and Winston, 1962. 278 pp.

c. Clements, Guy R., and Wilson, Levi T. *Analytical and Applied Mechanics.* (Third Edition.) New York: McGraw-Hill Book Company, 1951. 463 pp.

d. Synge, John L., and Griffith, B. A. *Principles of Mechanics.* (Third Edition.) New York: McGraw-Hill Book Company, 1959. 552 pp.

e. Slater, John C., and Frank, Nathaniel H. *Mechanics.* New York: McGraw-Hill Book Company, 1947. 297 pp.

f. Symon, Keith R. *Mechanics.* (Second Edition.) Reading, Mass.: Addison-Wesley Publishing Company, 1960. 557 pp.

g. Marion, Jerry B. *Classical Dynamics; Particles and Systems.* New York: Academic Press, 1965. 576 pp.

h. Greenwood, Donald. *Principles of Dynamics.* Englewood Cliffs, N. J.: Prentice-Hall, Inc., 1965. 518 pp.

See also Searle's *Experimental Harmonic Motion.*

See also "Resource Letter CM-1 on the Teaching of Angular Momentum and Rigid Body Motion," by John I. Shonle. *American Journal of Physics,* 33: 879-887, November 1965.

More advanced presentations include:

a. Becker, Robert A. *Introduction to Theoretical Mechanics.* New York: McGraw-Hill Book Company, 1954. 420 pp.

b. Coe, Carl J. *Theoretical Mechanics.* New York: The Macmillan Company, 1938. 555 pp.

c. Goldstein, Herbert. *Classical Mechanics.* Reading, Mass.: Addison-Wesley Publishing Company, 1950. 399 pp.

d. Milne, E. A. *Vectorial Mechanics.* New York: Interscience Publishers, 1948. 382 pp.

The author maintains that "vectors are not merely a pretty toy, suitable for elegant proofs of general theorems, but are a powerful weapon of workaday mathematical investigation."

For impacting bodies, elastic and inelastic, see:

Goldsmith, Werner. *Impact; The Theory and Physical Behaviour of Colliding Solids.* London: Edward Arnold and Company, 1960. 379 pp.

Gravity and friction may be included here, for they are universally encountered in dynamics. The classic presentation of the former is found in Newton's *Principia Mathematica:* [6]

Newton, Sir Isaac. *Mathematical Principles of Natural Philosophy,* translated into English by Andrew Motte in 1729; the translations revised, and supplied with an historical and explanatory appendix, by Florian Cajori. Berkeley, Cal.: University of California Press, 1934. 680 pp.

Newton's doctrine of gravitation is further expounded in:

Snow, A. J. *Matter and Gravity in Newton's Physical Philosophy.* London: Oxford University Press, 1926. 256 pp.

Elementary and advanced, respectively, are:

a. Gamow, George. *Gravity.* Garden City, N. Y.: Doubleday and Company, 1962. 157 pp.

b. Witten, Louis, editor. *Gravitation; An Introduction to Current Research.* New York: John Wiley and Sons, 1962. 481 pp.

Potential theory, a branch of mathematics having electrical as well as gravitational implications, is covered by:

a. MacMillan, William D. *The Theory of the Potential.* New York: McGraw-Hill Book Company, 1930. 469 pp.

---

6. See also J. Herivel, *The Background to Newton's Principia; A Study of Newton's Dynamical Researches in the Years 1664-84.* Oxford: At the Clarendon Press, 1965. 337 pp.

b. Ramsey, Arthur S. *An Introduction to the Theory of Newtonian Attraction.* Cambridge, England: At the University Press, 1940. 184 pp.

The contents are: (1) Preliminary mathematics; (2) Gravitational attraction and potential; (3) Attraction and potential at internal points; (4) Theorems of Laplace, Poisson and Gauss; (5) Green's theorem; (6) Harmonic functions; and (7) Attraction of ellipsoids.

c. Kellogg, Oliver D. *Foundations of Potential Theory.* Berlin: Springer, 1929. 384 pp.

Friction is treated in the following books:

a. Stanton, Thomas E. *Friction.* London: Longmans, Green and Company, 1923. 183 pp.

This collects widely scattered material on surface friction, rolling friction, viscosity, etc.

b. Gemant, Andrew. *Frictional Phenomena.* Brooklyn, N. Y.: Chemical Publishing Company, 1950. 497 pp.

The physics of friction is explained, in general and for various media such as gases, liquids, and solids.

c. Bowden, Frank P., and Tabor, D. *The Friction and Lubrication of Solids.* Oxford: At the Clarendon Press, 1950-1964. 2 vols. (parts).

See also Palmer's résumé.[7]

See also "Resource Letter F-1 on Friction," by Ernest Rabinowicz. *American Journal of Physics,* 31: 897-900, December 1963.

Fluid mechanics.

An historical and bibliographical treatment of the fluid concept is provided by:

Seeger, Raymond J. "On the Physics of Fluids." *American Journal of Physics,* 27: 377-387, September 1959.

Introductory textbooks, especially those designed for engineering instruction, are very numerous. Typical examples follow:

a. Binder, Raymond C. *Fluid Mechanics.* (Fourth Edition.) Englewood Cliffs, N. J.: Prentice-Hall, Inc., 1962. 453 pp.

b. Streeter, Victor L.[8] *Fluid Mechanics.* (Fourth Edition.) New York: McGraw-Hill Book Company, 1966. 705 pp.

---

7. F. Palmer, "What about Friction?" *American Journal of Physics,* 17: 181-187; 327-335; 336-342, April and September, 1949.
8. See also his *Handbook of Fluid Dynamics.* New York: McGraw-Hill Book Company, 1961. 1228 pp.

c. Vennard, John K. *Elementary Fluid Mechanics.* (Fourth Edition.) New York: John Wiley and Sons, 1961. 570 pp.

More advanced treatments of hydro- and aerodynamics (comprising fluid dynamics) include:

a. Lamb, Sir Horace. *Hydrodynamics.* (Sixth Edition.) Cambridge, England: At the University Press, 1932. 738 pp.
This standard comprehensive treatise is well documented.

b. Milne-Thomson, Louis M. *Theoretical Hydrodynamics.* (Fourth Edition.) New York: The Macmillan Company, 1960. 660 pp.

c. Milne-Thomson, Louis M. *Theoretical Aerodynamics.* (Second Edition.) New York: D. Van Nostrand Company, 1952. 414 pp.

d. Robertson, James M. *Hydrodynamics in Theory and Application.* Englewood Cliffs, N. J.: Prentice-Hall, Inc., 1965. 652 pp.

e. Howarth, L., editor. *Modern Developments in Fluid Dynamics*[9] *—High Speed Flow.* London: Oxford University Press, 1953. 2 vols.

f. Shapiro, Ascher H. *The Dynamics and Thermodynamics of Compressible Fluid Flow.* New York: Ronald Press Company, 1953-1954. 2 vols.

g. Brodkey, Robert S. *The Phenomena of Fluid Motions.* Reading, Mass.: Addison-Wesley Publishing Company, 1967. 737 pp.

The interaction of electrically-conducting fluids with a magnetic field is described in:

a. Jeffrey, A. *Magnetohydrodynamics.* New York: John Wiley and Sons, 1966. 252 pp.

b. Shercliff, J. A. *A Textbook of Magnetohydrodynamics.* New York: Pergamon Press, 1965. 265 pp.

c. Kulikovskiy, A. G., and Lyubimov, G. A. *Magnetohydrodynamics.* Reading, Mass.: Addison-Wesley Publishing Company, 1965. 216 pp.

d. Hughes, W. F., and Young, F. J. *Electromagnetodynamics of Fluids.* New York: John Wiley and Sons, 1966. 648 pp.

For ultrahigh velocity flow see:

Hayes, Wallace D., and Probstein, Ronald F. *Hypersonic Flow Theory.* (Second Edition.) Vol. 1: *Inviscid Flows.* New York: Academic Press, 1966. 602 pp.

For slow flow:

a. Scheidegger, Adrian E. *The Physics of Flow Through Porous Media.* (Revised Edition.) New York: The Macmillan Company, 1960. 313 pp.

---

9. Companion volumes to S. Goldstein's set (1938) in the same series.

b. Carman, P. C. *Flow of Gases Through Porous Media.* New York: Academic Press, 1956. 192 pp.

See also index under Aerodynamics and Air Flow.

Properties.

As the general properties of matter are fully discussed in most general textbooks[10] and compendia, this section will deal chiefly with elasticity, viscosity, and plasticity, comprising *rheology,* derived as follows:

> Founded in 1929 . . . the Society of Rheology was created to promote in all possible ways the study of the deformation and flow of matter. Prior to 1929, there was no single word adequate to define this particular field, including the subjects of elasticity, viscosity, and plasticity; the term "rheology" was therefore coined from the Greek "to flow" and "science" and was defined as "the science of the deformation and flow of matter." In recent years the word has become a standard part of the modern scientific vocabulary.[11]

Elasticity is reviewed on a physical basis, preparatory to mastery of its mathematical phases, in:

Southwell, R. V. *An Introduction to the Theory of Elasticity for Engineers and Physicists.* (Second Edition.) London: Oxford University Press, 1941. 509 pp.

Advanced theory is covered by:

a. Love, A. E. H. *A Treatise on the Mathematical Theory of Elasticity.* (Fourth Edition.) Cambridge, England: At the University Press, 1927. 643 pp.

This is the standard treatise on the subject, and is invaluable to physicists and engineers alike. (Dover Publications, Inc., has issued a reprint.)

b. Sokolnikoff, Ivan S. *Mathematical Theory of Elasticity.* (Second Edition.) New York: McGraw-Hill Book Company, 1956. 476 pp.

c. Timoshenko, Stephen, and Goodier, J. N. *Theory of Elasticity.* (Second Edition.) New York: McGraw-Hill Book Company, 1951. 506 pp.

d. Pearson, Carl E. *Theoretical Elasticity.* Cambridge, Mass.: Harvard University Press, 1959. 218 pp.

---

10. Notably those by Newman and Searle, and by Champion and Davy.
11. *Physics Today,* 4 (No. 10): 24, October 1951. See also the Society's *Transactions,* Vol. 1- 1957- New York: Interscience Publishers.

Photoelasticity renders elastic effects visible for practical and theoretical study, as Coker and Filon point out:

> The practical importance of Photo-Elasticity to the engineer can hardly be overrated. It provides him, as no other existing method does, with an immediate practical solution of fundamental problems concerning the stresses in the elements of structures and machines, which cannot be otherwise directly observed and which are usually beyond the reach of calculation. . . . But Photo-Elasticity has also its value for the pure Physicist, and it provides an additional means of exploring the interaction of molecules and atoms with radiation, a means to which little attention has hitherto been paid, and which should not be neglected, as it may throw much light upon the condition of matter in the solid state.[12]

Two excellent treatises are available:

a. Coker, Ernest G.; Filon, L. N. G.; and Jessop, H. T. *A Treatise on Photoelasticity.* (Second Edition.) Cambridge, England: At the University Press, 1957. 720 pp.

b. Frocht, Max M. *Photoelasticity.* New York: John Wiley and Sons, 1941-1948. 2 vols.

Engineering aspects are emphasized along theoretical and experimental lines.

See also Elasticity in the index.

Viscosity and plasticity receive comprehensive treatment in these surveys:

a. Nádai, A. *Theory of Flow and Fracture of Solids.* New York: McGraw-Hill Book Company, 1950-1963. 2 vols.

Vol. 1 is a second edition of his *Plasticity* (1931).

b. Hatschek, Emil. *The Viscosity of Liquids.* London: G. Bell and Sons, 1928. 239 pp.

c. Reiner, Markus. *Deformation, Strain and Flow; An Elementary Introduction to Rheology.* (Second Edition.) New York: Interscience Publishers, 1960. 347 pp.

d. Houwink, R. *Elasticity, Plasticity and Structure of Matter.* (Second Edition.) New York: Dover Publications, 1959. 368 pp.

e. Fredrickson, A. G. *Principles and Applications of Rheology.* Englewood Cliffs, N. J.: Prentice-Hall, Inc., 1964. 326 pp.

f. Eirich, Frederick R., editor. *Rheology; Theory and Applications.* New York: Academic Press, 1956-1967. 4 vols.

Further references may be found in:

Kececioglu, Dimitri. *Bibliography of Plasticity.* New York: American Society of Mechanical Engineers, 1950. 191 pp.

---

12. P. xiv of their *Treatise* . . .

Covering both theory and application, this comprehensive list of 40 books and 1845 articles is arranged chronologically (1837-1949), with author and subject indexes.

See also Viscosity in the index.

Micromeritics is the name applied to the technology of fine particles, larger than colloids and up to one inch diameter:

a. Dalla Valle, J. M. *Micromeritics; The Technology of Fine Particles.* (Second Edition.) New York: Pitman Publishing Corporation, 1948. 555 pp.
The author states (p. vii): "There is no phase of engineering or applied science where some knowledge of particular technology is not required."
Chapters include: Dynamics of small particles; Electrical, optical and sonic properties; Thermodynamics of particles; Capillarity.

b. Cadle, R. D. *Particle Size Determination.* New York: Interscience Publishers, 1955. 303 pp.

c. Herdan, Gustav. *Small Particle Statistics.* (Second Edition.) New York: Academic Press, 1960. 418 pp.

d. Orr, Clyde. *Particulate Technology.* New York: The Macmillan Company, 1966. 562 pp.

High and low pressure effects.

During the course of his half-century of outstanding work in high-pressure physics, this Nobel laureate had prepared a lucid summary of his own and other contributions:

Bridgman, Percy W. *The Physics of High Pressure.* (New Impression with Supplement.) London: G. Bell and Sons, 1949. 445 pp.
An historical introduction and copious bibliographical references were included.
See also Bridgman's *Collected Experimental Papers* for the complete record of his accomplishments.

A section of the following bibliography indexes Bridgman's writings:

Annotated Bibliography on High-Pressure Technology, compiled by Alexander Zeitlin. Washington, D. C.: Butterworths Scientific Publications, 1964. 290 pp.
Published jointly with American Society of Mechanical Engineers.

Subsequent developments are presented in:

a. Paul, William, and Warschauer, Douglas M., editors. *Solids Under Pressure.* New York: McGraw-Hill Book Company, 1963. 478 pp.
These fifteen articles constitute a memorial tribute to Bridgman.

b. Bradley, R. S. *High Pressure Physics and Chemistry.* New York: Academic Press, 1963. 2 vols.

c. Wentorf, R. H., Jr., editor. *Modern Very High Pressure Techniques.* Washington, D. C.: Butterworths Scientific Publications, 1962. 232 pp.

d. Tomizuka, C. T., and Emrick, R. M., editors. *Physics of Solids at High Pressures.* New York: Academic Press, 1965. 595 pp. These international conference proceedings cover electronic properties, magnetism, phase transformations and lattice properties.

Low pressure (or high vacuum) practice is covered by numerous books, among which are:

a. Dushman, Saul. *Scientific Foundations of Vacuum Technique,* edited by J. M. Lafferty. (Second Edition.) New York: John Wiley and Sons, 1962. 806 pp.

b. Yarwood, John. *High Vacuum Technique: Theory, Practice, and Properties of Materials.* (Fourth Edition.) London: Chapman and Hall, Ltd., 1967. 274 pp.

c. Roberts, Richard W., and Vanderslice, Thomas A. *Ultrahigh Vacuum and Its Applications.* Englewood Cliffs, N. J.: Prentice-Hall, Inc., 1963. 199 pp.

d. Lewin, Gerhard. *Fundamentals of Vacuum Science and Technology.* New York: McGraw-Hill Book Company, 1965. 248 pp.

e. Guthrie, Andrew. *Vacuum Technology.* New York: John Wiley and Sons, 1963. 532 pp.

There is also a handbook:

Beck, Arnold H. W., editor. *Handbook of Vacuum Physics.* Oxford: Pergamon Press, 1964-1966. 3 vols. in 13 parts. Volume contents: 1, Gases and vacua; 2, Physical electronics; and 3, Technology.

Of interest are the transactions of the American Vacuum Society, published since 1954. Beginning 1964 these National Vacuum Symposia will be reported in the *Journal of Vacuum Science and Technology* instead of separately.

Transactions of the third International Vacuum Congress (1965) were published by Pergamon Press, 1967, in three parts.

Applied mechanics.

Some examples of the engineering mechanics books omitted from general treatment may be cited:

a. Timoshenko, Stephen, and Young, D. H. *Engineering Me-*

chanics. (Fourth Edition.) New York: McGraw-Hill Book Company, 1956. 534 pp.

b. Singer, Ferdinand L. *Engineering Mechanics.* (Second Edition.) New York: Harper and Row, 1954. 525 pp.

c. Seely, Fred B.; Ensign, N. E.; and Jones, P. G. *Analytical Mechanics for Engineers.* (Fifth Edition.) New York: John Wiley and Sons, 1958. 475 pp.

d. Shames, Irving H. *Engineering Mechanics; Statics and Dynamics.* (Second Edition.) Englewood Cliffs, N. J.: Prentice-Hall, Inc., 1966. 2 vols. in 1.

The gyroscope is an important application of mechanics, and is best introduced by comparison with tops: [13]

Crabtree, Harold. *An Elementary Treatment of the Theory of Spinning Tops and Gyroscopic Motion.* (Second Edition.) London: Longmans, Green and Company, 1914. 193 pp.
Rotational dynamics are studied in popular style without sacrifice of scientific accuracy.

More advanced theory and applications are found in:

a. Richardson, K. I. T. *The Gyroscope Applied.* New York: Philosophical Library, 1954. 384 pp.

b. Cochin, Ira. *Analysis and Design of the Gyroscope for Inertial Guidance.* New York: John Wiley and Sons, 1963. 156 pp.
Covers gyrodynamics, suspension, and drift.

c. Arnold, Ronald N., and Maunder, Leonard. *Gyrodynamics and Its Engineering Applications.* New York: Academic Press, 1961. 484 pp.

Vibration prevention is another field of applied mechanics, treated in:

a. Den Hartog, J. P. *Mechanical Vibrations.* (Fourth Edition.) New York: McGraw-Hill Book Company, 1956. 436 pp.

b. Thomson, William T. *Vibration Theory and Applications.* Englewood Cliffs, N. J.: Prentice-Hall, Inc., 1965. 384 pp.

c. Church, Austin H. *Mechanical Vibrations.* (Second Edition.) New York: John Wiley and Sons, 1963. 432 pp.

d. Wilson, W. Ker. *Practical Solution of Torsional Vibration Problems.* (Third Edition.) Vols. 1-3. New York: John Wiley and Sons, 1956-1965. (In Process.)

---

13. There is even a four-volume handbook: F. Klein and A. Sommerfeld, *Theorie des Kreisels* (Leipzig, 1897-1910).

A third major application of the principles of mechanics is ballistics, which has two subdivisions:

> *Ballistics* is the art of hurling a projectile. It is ordinarily divided into two branches, *interior ballistics,* which deals with the projectile and its accessories while it is within the weapon, and *exterior ballistics,* which deals with the projectile from the gun to the target.[14]

Varied books on ballistics:

a. McShane, Edward J.; Kelley, John L.; and Reno, Franklin V. *Exterior Ballistics.* Denver: University of Denver Press, 1953. 834 pp.

b. Bliss, Gilbert A. *Mathematics for Exterior Ballistics.* New York: John Wiley and Sons, 1944. 128 pp.

c. Davis, Leverett, Jr., *et al. Exterior Balistics of Rockets.* Princeton, N. J.: D. Van Nostrand Company, 1958. 457 pp.

d. Corner, J. *Theory of the Interior Ballistics of Guns.* New York: John Wiley and Sons, 1950. 443 pp.

Closely related are the dynamics of space flight in astronautics:

a. Thomson, William T. *Introduction to Space Dynamics.* New York: John Wiley and Sons, 1961. 317 pp.
Physics applicable to ballistic missiles, satellites, probes and rockets.

b. Roy, Archie E. *The Foundations of Astrodynamics.* New York: The Macmillan Company, 1965. 385 pp.

c. Berman, Arthur I. *The Physical Principles of Astronautics; Fundamentals of Dynamical Astronomy and Space Flight.* New York: John Wiley and Sons, 1961. 350 pp.

d. Koelle, Heinz H., editor. *Handbook of Astronautical Engineering.* New York: McGraw-Hill Book Company, 1961. 1814 pp.

e. Baker, Robert M. L., and Makemson, Maude W. *An Introduction to Astrodynamics.* New York: Academic Press, 1967. 439 pp.

See also "Resource Letter SO-1 on the Kinematics and Dynamics of Satellite Orbits," by Leon Blitzer. *American Journal of Physics,* 31: 233-236, April 1963.

For bibliographical aid see:

a. Ordway, Frederick I. *Annotated Bibliography of Space Science and Technology: A History of Astronautical Book Literature—1931 through 1961.* (Third Edition.) Washington, D. C.: ARFOR Publications, 1962. 77 pp.

b. Benton, Mildred C. *The Literature of Space Science and Ex-*

---

14. C. S. Robinson, *Thermodynamics of Firearms,* p. 2. New York McGraw-Hill Book Company, 1943.

ploration. Washington, D. C.: U. S. Naval Research Laboratory, 1958. 264 pp.
This covers books, articles and reports from 1903 through mid-1958.

c. Fry, Bernard M., and Mohrhardt, Foster E., editors. *A Guide to Information Sources in Space Science and Technology.* New York: Interscience Publishers, 1963. 579 pp.

See also index under Astrophysics.

3—Heat

General.

Comprehensive treatments of the form of energy called heat, and its relationship with work (thermodynamics), are the following:

a. Saha, M. N., and Srivastava, B. N. *A Treatise on Heat.* (Fourth Edition.) Allahabad and Calcutta: The Indian Press, 1958. 275 pp.
This is well known for its clarity of presentation.

b. Becker, Richard. *Theory of Heat.* (Second Edition.) Berlin: Springer, 1967. 380 pp.

c. Planck, Max. *Treatise on Thermodynamics.* (Third Edition.) London: Longmans, Green and Company, 1927. 297 pp.

d. Hatsopoulos, George N., and Keenan, Joseph H. *Principles of General Thermodynamics.* New York: John Wiley and Sons, 1965. 788 pp.

The textbooks in this field are very numerous, especially those on applied thermodynamics intended for engineering school use. Some which are oriented towards physics follow:

a. Zemansky, Mark W. *Heat and Thermodynamics; An Intermediate Textbook for Students of Physics, Chemistry, and Engineering.* (Fourth Edition.) New York: McGraw-Hill Book Company, 1957. 484 pp.
Fundamental principles and applications are clearly developed, with many diagrammatic representations of systems and thermodynamic surfaces.

b. Morse, Philip M. *Thermal Physics.* New York: W. A. Benjamin, Inc., 1964. 455 pp.
Thermodynamics, kinetic theory and statistical mechanics are covered, in this book and the next one.

c. King, Allen L. *Thermophysics.* San Francisco: W. H. Freeman and Company, 1962. 369 pp.

d. Roberts, John K. *Heat and Thermodynamics,* revised by A. R.

Miller. (Fifth Edition.) New York: Interscience Publishers, 1960. 619 pp.

e. Haar, Dirk ter, and Wergeland, H. N. S. *Elements of Thermodynamics.* Reading, Mass.: Addison-Wesley Publishing Company, 1966. 160 pp.

f. Guggenheim, Edward A. *Thermodynamics: An Advanced Treatment for Chemists and Physicists.* (Fifth Edition.) Amsterdam: North-Holland Publishing Company, 1967. 390 pp.

General backgrounds are sketched in different manners by:

a. Bridgman, Percy W. *The Nature of Thermodynamics.* Cambridge, Mass.: Harvard University Press, 1941. 229 pp.
The author made one of his characteristic operational analyses, "in such a form as to bring out the role and the consequences of the verbal requirements which have entered into forming the concepts." [15]

b. McKie, Douglas, and Heathcote, Niels H. de V. *The Discovery of Specific and Latent Heats.* London: Edward Arnold and Company, 1935. 155 pp.
In the preface of this scholarly presentation based on original sources, E. N. da C. Andrade states (p. 4):

> The subject with which the book deals, under its unpretentious title, is no less than the foundation of the modern science of heat, which may be said to have originated when a really clear distinction was made between heat and temperature.

Formulas for fundamental and derived entities in thermodynamics are conveniently gathered in:

Bridgman, Percy W. *A Condensed Collection of Thermodynamic Formulas.* Cambridge, Mass: Harvard University Press, 1925. 34 pp.

Diagrams in the form of entropy charts are found in:

Gourdet, G., and Proust, A. *Les Diagrammes Thermodynamiques.* Paris: Dunod, 1950. 2 vols.

Current periodical indexing media described in Chapter II may be supplemented for early material by:

Tuckerman, Alfred. *Index to the Literature of Thermodynamics.* Washington, D. C.: Smithsonian Institution, 1890. 239 pp. (Smithsonian Miscellaneous Collections, Vol. 34, No. 741.)

For thermophysical references since 1940 see two sets compiled at Purdue University by Y. S. Touloukian, via author index.

---

15. *Op. cit.*, p. xii.

Kinetic theory of gases.

The conception of gases as bodies of particles in constant rebounding motion leads to fundamental laws expressed in:

a. Jeans, Sir James H. *An Introduction to the Kinetic Theory of Gases.* New York: The Macmillan Company, 1940. 311 pp. Physical rather than mathematical aspects are emphasized.

b. Kennard, E. H. *Kinetic Theory of Gases.* New York: McGraw-Hill Book Company, 1938. 483 pp.

c. Present, R. D. *Kinetic Theory of Gases.* New York: McGraw-Hill Book Company, 1958. 280 pp.

See also index under Plasma physics, and Collision phenomena.

Combustion processes are described in:

a. Lewis, Bernard, and Elbe, Guenther von. *Combustion, Flames and Explosions of Gases.* (Second Edition.) New York: Academic Press, 1961. 731 pp.

b. *Symposium (International) on Combustion.* "Papers of the first symposium were published in *Industrial and Engineering Chemistry,* Vol. 20, pp. 998-1057, 1928; and of the second in *Chemical Reviews,* Vol. 21, pp. 209-460, 1937, and Vol. 22, pp. 1-310, 1938." Thereafter, issued in bound form by various publishers, presently by The Combustion Institute, Pittsburgh. (The volume issued for the tenth symposium held at Cambridge University in 1964 includes cumulative author and subject indexes to all.)

Various aspects of flames are described in:

a. Gaydon, Alfred G., and Wolfhard, H. G. *Flames; Their Structure, Radiation and Temperature.* (Second Edition.) New York: The Macmillan Company, 1960. 383 pp.

b. Gaydon, Alfred G. *The Spectroscopy of Flames.* New York: John Wiley and Sons, 1957. 279 pp.

c. Weinberg, F. J. *Optics of Flames.* London: Butterworths Scientific Publications, 1963. 251 pp.

d. Fristrom, R. M., and Westenberg, A. A. *Flame Structure.* New York: McGraw-Hill Book Company, 1965. 424 pp.

Heat flow.

Theory and applications are covered by the treatise:

Jakob, Max, and Kezios, S. P. *Heat Transfer.* New York: John Wiley and Sons, 1949-1957. 2 vols.

For conduction processes, one may consult:

a. Carslaw, Horatio S., and Jaeger, J. C. *Conduction of Heat in Soilds.* (Second Edition.) Oxford: At the Clarendon Press, 1959. 510 pp.

b. Ingersoll, Leonard, *et al. Heat Conduction, with Engineering, Geological, and Other Applications.* (Revised Edition.) Madison, Wisc.: The University of Wisconsin Press, 1954. 325 pp.

Heat transfer in the engineering sense is thoroughly treated in a survey sponsored by the National Research Council:

McAdams, William H. *Heat Transmission.* (Third Edition.) New York: McGraw-Hill Book Company, 1954. 532 pp.

High and low temperature effects.

Extremes of temperature are interestingly treated in one of the Momentum Books published for the Commission on College Physics:

Zemansky, Mark W. *Temperatures Very Low and Very High.* Princeton, N. J.: D. Van Nostrand Company, 1964. 127 pp.

Cryophysics and *cryogenics* are now associated with research at low temperatures, involving property study thereunder as well as production techniques.

One might first read for introductory purposes:

a. Jackson, L. C. *Low Temperature Physics.* (Fifth Edition.) London: Methuen and Company, 1962. 158 pp.

b. Mendelssohn, K. *The Quest for Absolute Zero: The Meaning of Low Temperature Physics.* New York: McGraw-Hill Book Company, 1966. 256 pp.

There is a comprehensive bibliography:

Codlin, Ellen M. *Cryogenics and Refrigeration: A Bibliographical Guide.* New York: Plenum Press, 1968. 288 pp.

Low temperature work is reviewed in:

a. Squire, Charles F. *Low Temperature Physics.* New York: McGraw-Hill Book Company, 1953. 244 pp.

b. Ruhemann, M., and Ruhemann, B. *Low Temperature Physics.* Cambridge, England: At the University Press, 1937. 313 pp. An extensive bibliography appears on pp. 291-313.

c. Burton, Eli F.; Smith, H. G.; and Wilhelm, J. O. *Phenomena at the Temperature of Liquid Helium.* New York: Reinhold Publishing Corporation, 1940. 362 pp.

d. Rosenberg, H. M. *Low Temperature Solid State Physics.* Oxford: At the Clarendon Press, 1963. 420 pp.

e. Vance, Robert W., editor. *Cryogenic Technology.* New York: John Wiley and Sons, 1963. 585 pp.

Some books on superfluidity:

a. London, Fritz. *Superfluids.* (Second Edition.) With a new epilogue, Theoretical Developments 1950-1960, by M. J. Buckingham. New York: Dover Publications, 1961-1964. 2 vols.
Macroscopic theories of superconductivity and superfluid helium are developed.

b. Khalatnikov, Isaak M. *Introduction to the Theory of Superfluidity.* New York: W. A. Benjamin, Inc., 1965. 206 pp.

c. Lane, Cecil T. *Superfluid Physics.* New York: McGraw-Hill Book Company, 1962. 226 pp.

d. Atkins, K. R. *Liquid Helium.* Cambridge, England: At the University Press, 1959. 312 pp.

e. Wilks, J. *The Properties of Liquid and Solid Helium.* London: Oxford University Press, 1967. 703 pp.

See also "Resource Letter LH-1 on Liquid Helium," by C. T. Lane. *American Journal of Physics,* 35: 367-375, May 1967.

A special aspect, namely the remarkably high electrical conductivity encountered in the neighborhood of absolute zero (−273° C), is described in:

a. Shoenberg, David. *Superconductivity.* (Second Edition.) Cambridge, England: At the University Press, 1952. 256 pp.

b. Rickayzen, G. *Theory of Superconductivity.* New York: Interscience Publishers, 1965. 483 pp.

c. Blatt, John M. *Theory of Superconductivity.* New York: Academic Press, 1964. 486 pp.
A fifty-page bibliography is appended to this survey.

See also "Resource Letter Scy-1 on Superconductivity," by D. M. Ginsberg. *American Journal of Physics,* 32: 85-89, February 1964.

Low-temperature magnetic effects appear in:

a. Casimir, H. B. G. *Magnetism and Very Low Temperatures.* Cambridge, England: At the University Press, 1940. 93 pp.

b. Garrett, C. G. B. *Magnetic Cooling.* New York: John Wiley and Sons, 1954. 110 pp.

See also index under Cryophysics.

For high temperatures, especially their measurement by instruments known as pyrometers, we have available:

a. Burgess, G. K., and LeChatelier, H. *The Measurement of High Temperatures.* (Third Edition.) New York: John Wiley and Sons, 1912. 510 pp.
Although old, this is still important as a well-documented survey from historical, theoretical and practical viewpoints.

b. Wood, William P., and Cork, James M. *Pyrometry.* (Second Edition.) New York: McGraw-Hill Book Company, 1941. 263 pp. This is designed as a textbook and a practical treatise as well.

c. Harrison, Thomas R.[16] *Radiation Pyrometry and Its Underlying Principles of Radiant Heat Transfer.* New York: John Wiley and Sons, 1960. 234 pp.

d. Campbell, I. E., editor. *High-Temperature Technology.* New York: John Wiley and Sons, 1956. 526 pp.
Methods for producing and measuring high temperatures are described in relation to refractories.

e. Kingery, W. D. *Property Measurements at High Temperatures: Factors Affecting and Methods of Measuring Material Properties at Temperatures Above 1400° C (2550° F).* New York: John Wiley and Sons, 1959. 416 pp.

See also Temperature in the index.

Applied thermodynamics.

An interesting translation of a French classic on heat engines is:

Carnot, Sadi. *Reflections on the Motive Power of Heat, and on Machines fitted to Develop this Power.* Translated by R. H. Thurston. New York: American Society of Mechanical Engineers, 1943. 107 pp.
The 1824 work is here presented in superb format, accompanied by a biography of Carnot.

Clean-cut diagrams in color are a noteworthy feature of:

Wrangham, D. A. *The Theory and Practice of Heat Engines.* Cambridge, England: At the University Press, 1942. 756 pp.

For power applications, consult Kent's and Marks' handbooks of mechanical engineering, and comprehensive textbooks such as:

Solberg, H. L.; Cromer, O. C.; and Spalding, A. R. *Thermal Engineering.* New York: John Wiley and Sons, 1960. 649 pp.

4—Sound

General.

The outstanding classic in advanced acoustical theory is the following treatise:

Rayleigh, John W. S. *The Theory of Sound.* (Second Edition.) London: The Macmillan Company, 1926. 2 vols.

---

16. Also co-author of the classic *Pyrometric Practice.* 1921. 326 pp. (U. S. National Bureau of Standards Technical Paper No. 170.)

There is also a single volume photo-reproduction (New York: Dover Publications, 1945) that incorporates a historical introduction by Robert B. Lindsay, who states:

> It is scarcely an exaggeration to say that there is no vibrating system likely to be encountered in practice which cannot be tackled successfully by the methods set forth in the first ten chapters of Rayleigh's treatise. Even the worker in the field of non-linear systems, a department of increasing practical importance in modern vibration theory, will find useful basic hints in Rayleigh.

Based largely on the foregoing is the equally well known:

Lamb, Sir Horace. *The Dynamical Theory of Sound.* (Second Edition.) London: Edward Arnold and Company, 1925. 307 pp.

Mathematical treatments appear also in the following advanced textbooks:

a. Morse, Philip M. *Vibration and Sound.* (Second Edition.) New York: McGraw-Hill Book Company, 1948. 468 pp.

b. Stewart, George W., and Lindsay, Robert B. *Acoustics; A Text on Theory and Applications.* New York: D. Van Nostrand Company, 1930. 358 pp.

c. Stephens, Raymond W. B., and Bate, A. E. *Acoustics and Vibrational Physics.* (Second Edition.) London: Edward Arnold, Ltd., 1966. 818 pp.

d. Bickley, W. G., and Talbot, A. *An Introduction to the Theory of Vibrating Systems.* Oxford: At the Clarendon Press, 1961. 238 pp.

e. Lindsay, Robert B. *Mechanical Radiation.* New York: McGraw-Hill Book Company, 1960. 415 pp.

Uses mechanical radiation to illustrate all kinds of wave propagation. Topics include: wave motion and properties; harmonic analysis; waves in strings, membranes, rods, plates, fluids, etc.; acoustic radiation and the properties of matter.

f. Towne, Dudley H. *Wave Phenomena.* Reading, Mass.: Addison-Wesley Publishing Company, 1967. 482 pp.

A unified treatment of mechanical, acoustic and electromagnetic waves.

There is a comprehensive reference set:

Mason, Warren P., editor. *Physical Acoustics; Principles and Methods.* New York: Academic Press, 1964- 6 vols. (In Process.)

Some introductory textbooks on acoustics follow:

a. Wood, Albert B. *A Textbook on Sound.* (Third Edition.) New York: The Macmillan Company, 1955. 610 pp.

This is an excellent account of the physics of mechanical vibrations, both audible and inaudible. Many interesting applications are described, such as echo sounding, sound ranging, sound motion pictures, etc. Sound measurement, analysis, and recording are well summarized.

b. Richardson, Edward G. *Sound; A Physical Textbook.* (Fifth Edition.) London: Edward Arnold and Company, 1953. 352 pp.

c. Randall, Robert H. *An Introduction to Acoustics.* Cambridge, Mass.: Addison-Wesley Press, 1951. 340 pp.

d. Hastings, Russell B. *The Physics of Sound.* St. Paul, Minn.: Bruce Publishing Company, 1960. 259 pp.
Forty experiments are included.

e. Watson, Floyd R. *Sound; An Elementary Textbook on the Science of Sound and the Phenomena of Hearing.* New York: John Wiley and Sons, 1935. 219 pp.
"Reference books on sound" (pp. vii-ix) is an excellent annotated bibliography, and the last chapter outlines 19 experiments on sound.

f. Kinsler, Lawrence E., and Frey, Austin R. *Fundamentals of Acoustics.* (Second Edition.) New York: John Wiley and Sons, 1962. 524 pp.

For measurement, see Beranek's *Acoustic Measurements* in conjunction with:

Rao, V. V. L. *The Decibel Notation; Its Application to Radio and Acoustics.* Brooklyn, N. Y.: Chemical Publishing Company, 1946. 179 pp.

Physiological acoustics.

The word "sound" may connote the sensation of hearing, as well as the vibratory disturbances capable of producing it. The classic discussion of this interrelationship is:

Helmholtz, Hermann L. F. von. *On the Sensations of Tone as a Physiological Basis for the Theory of Music*, translated and revised by Alexander J. Ellis. (Fourth Edition.) London: Longmans, Green and Company, 1912. 575 pp.
Its three parts are: (1) On the composition of vibrations; (2) On the interruptions of harmony; and (3) The relationship of musical tones.

More recent presentations, developed from physical and physiological backgrounds, respectively, are:

a. Fletcher, Harvey. *Speech and Hearing in Communication.* New York: D. Van Nostrand Company, 1953. 461 pp.
This brings to date the author's earlier *Speech and Hearing.* Some chapter titles are: The speech sounds of English; The speaking

mechanism; Characteristics of speech waves; Noise; Mechanism of hearing; Minimum perceptual changes in frequency and sound pressure level; Loudness.

b. Stevens, Stanley S., and Davis, Hallowell. *Hearing; Its Psychology and Physiology.*[17] New York: John Wiley and Sons, 1938. 489 pp.

Very useful are the glossary (pp. 449-456), and the bibliography (pp. 457-472). Its joint authorship by a psychologist and physiologist is aptly defended by their comment (p. xi):

> It is the characteristic of the science of audition that it ignores the traditional boundaries between the sciences. None of the traditional disciplines nor any of the academic departments of the modern university can claim audition exclusively as its own. The mystery of the ear inspires the psychologist, the physiologist, the otologist, and the physicist alike.

Over 10,000 titles are listed in:

Harvard University Psycho-Acoustic Laboratory. *Bibliography on Hearing.* Prepared by the Laboratory; S. S. Stevens, director; J. G. C. Loring, compiler; Dorothy Cohen, technical editor. Cambridge, Mass.: Harvard University Press, 1955. 599 pp.

This is an enlargement of a bibliography by G. A. Miller, *et al.*

For the mechanics of the ear see:

Littler, T. S. *Physics of the Ear.* Oxford, New York, etc.: Pergamon Press, 1965. 378 pp.

Musical acoustics from the physicist's viewpoint is the subject of:

a. Jeans, Sir James H. *Science and Music.* New York: The Macmillan Company, 1937. 258 pp.

The style of this book is pleasantly non-technical.

b. Wood, Alexander. *The Physics of Music.* Cleveland, Ohio: The Sherwood Press, 1944. 255 pp.

This was reissued by Dover Publications in 1956. The author describes the "very interesting borderland between physics and music."

c. Culver, Charles A. *Musical Acoustics.* (Fourth Edition.) New York: McGraw-Hill Book Company, 1956. 297 pp.

d. Olson, Harry F. *Music, Physics and Engineering.* (Second Edition.) New York: Dover Publications, 1967. 460 pp.

The relation of the physical principles of sound to music from the musician's standpoint is delineated in:

---

17. See also S. S. Stevens and F. Warshofsky, *Sound and Hearing.* New York: Time, Inc., 1965. 200 pp. (Life Science Library.)

Bartholomew, Wilmer T. *Acoustics of Music.* New York: Prentice-Hall, Inc., 1942. 242 pp.
An extensive bibliography is appended (pp. 231-238). The author asserts (p. vii):

> The Taj Mahal, that dream of architectural loveliness, seems to our entranced vision far removed from such mundane things as the measurement of angles, the stresses and strains of building materials, and the chemistry of pigments. But is it not related to these matters as the harvest is to the seed? Beethoven's Ninth Symphony, that last mighty dream of a stormy soul, or *L'après midi,* that more delicate dream of fragile beauty, seems to our transported ears to have no connection with such prosaic things as the compressibility of air, the reflection of sound waves by walls, or the mathematics of Fourier analysis. But there is a connection, unrecognized and scorned though it be by many.

Diametrically opposed to musical tones are the discordant sounds treated in:

a. McLachlan, Norman W. *Noise; A Comprehensive Survey from Every Point of View.* London: Oxford University Press, 1935. 148 pp.
It discusses the origin, measurement, effects and reduction of various types of noise, and lists further references on pp. 139-145.

b. MacDonald, D. K. C. *Noise and Fluctuations: An Introduction.* New York: John Wiley and Sons, 1962. 118 pp.
As Lindsay states:

> Noise in acoustics is often defined as unwanted sound, but this is a far from satisfactory definition. Much more relevant is the viewpoint according to which noise is a random series of tones of a wide range of frequencies having no regular connection with each other. From this standpoint, there is a close analogy between acoustical noise and fluctuation phenomena.[18]

c. Beranek, Leo L. *Noise Reduction.* New York: McGraw-Hill Book Company, 1960. 752 pp.

d. Harris, Cyril M., editor. *Handbook of Noise Control.* New York: McGraw-Hill Book Company, 1957. 1184 pp.
Noise creation, measurement and reduction are fully covered by a staff of specialists.

See also index under Noise, electrical.

Ultrasonics.

Sound waves become inaudible to the average individual when of higher frequency than twenty kilocycles. Introductory acquaintance

18. R. B. Lindsay, in review. *Physics Today,* 16 (No. 9): 74, September 1963.

with ultrasonic phenomena is provided by Hubbard's survey [19] and by:

Wood, Robert Williams. *Supersonics; The Science of Inaudible Sounds.* Providence, R. I.: Brown University, 1939. (Reprinted 1948 with sup. bibl.) 164 pp.

Reviewing it in *Physics Today,* 2 (No. 2): 29, February 1949, Elias Klein states:

> Nowadays, supersonics pertains to speed above the normal sound velocity in air, while ultrasonics deals with high frequency inaudible sound waves. Therefore, the title of this book may be somewhat confusing, but its contents reveal clearly the early developments of ultrasound. . . . For essential background material in ultrasonics, this book remains an excellent primer, in spite of the outstanding technological advances which have been made in this field during the past ten years.

More extensive treatments are as follows:

a. Bergmann, Ludwig. *Der Ultraschall und Seine Anwendung in Wissenschaft und Technik.* (Sechste Auflage.) Stuttgart: S. Hirzel, 1954. 1114 pp.

The publisher issued a sixty-six page *Nachtrag zum Literaturverzeichnis* of this edition in 1957. This well-documented treatise, covering generation, detection, measurement, and applications, is also available in an earlier and less comprehensive English edition. (London: G. Bell and Sons, 1938. 264 pp.)

b. Carlin, Benson. *Ultrasonics.* (Second Edition.) New York: McGraw-Hill Book Company, 1960. 309 pp.

c. Richardson, Edward G. *Ultrasonic Physics,* edited by A. E. Brown. (Second Edition.) New York: American Elsevier Publishing Company, 1962. 313 pp.

This features the ultrasonic interferometer as a precision tool in physics laboratories.

d. Vigoreux, Paul. *Ultrasonics.* New York: John Wiley and Sons, 1951. 163 pp.

e. Blitz, Jack. *Fundamentals of Ultrasonics.* (Second Edition.) London: Butterworths Scientific Publications, 1967. 220 pp.

f. Goldman, Richard G. *Ultrasonic Technology.* New York: Reinhold Publishing Corporation, 1962. 320 pp.

Further references may be found in:

Curry, Beth, *et al. Bibliography: Supersonics or Ultrasonics, 1926 to 1949, with Supplement to 1950.* Stillwater, Okla.: Oklahoma Agricultural and Mechanical College, 1951. 277 pp.

---

19. J. C. Hubbard, "Ultrasonics—A Survey." *American Journal of Physics,* 8: 207-221, August 1940.

Applied acoustics.

Useful reference compilations on all phases of modern acoustics are:

a. Olson, Harry F. *Acoustical Engineering.* Princeton, N. J.: D. Van Nostrand Company, 1957. 718 pp.
This is virtually a third edition of his *Elements of Acoustical Engineering,* long a standard work.

b. Richardson, Edward G., and Meyer, E., editors. *Technical Aspects of Sound.* New York: American Elsevier Publishing Company, 1953-1962. 3 vols.
Volume titles of this collective survey are: (Vol. 1) Sonic range and airborne sound; (Vol. 2) Ultrasonic range; underwater acoustics; and (Vol. 3) Recent developments in acoustics.

"Analysis, testing and processing of materials and products by the use of mechanical vibratory energy" is the subject of:

Hueter, Theodor F., and Bolt, Richard H. *Sonics.* New York: John Wiley and Sons, 1955. 456 pp.

Undersea techniques are found in:

a. Tolstoy, I., and Clay, C. S. *Ocean Acoustics; Theory and Experiment in Underwater Sound.* New York: McGraw-Hill Book Company, 1966. 293 pp.

b. Albers, V. M., editor. *Underwater Acoustics.* New York: Plenum Press, 1963-1967. 2 vols.
These are NATO study institute proceedings.

Of architectural interest are:

a. Knudsen, Vern O., and Harris, C. M. *Acoustical Designing in Architecture.* New York: John Wiley and Sons, 1950. 457 pp.

b. Watson, Floyd R. *Acoustics of Buildings, including Acoustics of Auditoriums and Soundproofing of Rooms.* (Third Edition.) New York: John Wiley and Sons, 1941. 171 pp.

c. Beranek, Leo L. *Music, Acoustics and Architecture.* New York: John Wiley and Sons, 1962. 586 pp.

Some books on sound recording:

a. Pear, C. B. *Magnetic Recording in Science and Industry.* New York: Reinhold Publishing Corporation, 1967. 453 pp.

b. Read, Oliver. *The Recording and Reproduction of Sound.* (Second Edition.) Indianapolis, Ind.: Howard W. Sams and Company, 1952. 790 pp.

c. Mee, C. D. *The Physics of Magnetic Recording.* New York: Interscience Publishers, 1964. 269 pp.

d. Briggs, Gilbert A. *Sound Reproduction.* (Third Edition.) New York: Herman and Stephens, 1956. 368 pp.

5—Light

General.

An enjoyable introduction is provided by:

a. Tolansky, Samuel. *Curiosities of Light Rays and Light Waves.* New York: American Elsevier Publishing Company, 1965. 109 pp.

b. Tolansky, Samuel. *Optical Illusions.* Oxford: Pergamon Press; New York: The Macmillan Company, 1964. 156 pp.

Some of the headings are: Illusion in nature; Convergence-divergence; Irradiation illusions; Illusions produced by hatched lines; Illusions involving oscillation of attention; and Illusions due to instrumentation.

c. Cagnet, Michael; Françon, Maurice; and Thrierr, Jean Claude. *Atlas of Optical Phenomena.* Berlin: Springer; Englewood Cliffs, N. J.: Prentice-Hall, Inc., 1962. 94 pp. (45 pp. of plates.)

In his book review, Ballard states:

> Here is something that should warm the heart of every optics teacher and serve valiantly to better instruct his students. At last there is available . . . a series of illustrations of important phenomena in conventional geometrical and physical optics. To be sure, the usual textbooks do their best to show pictures of diffraction patterns, interference fringes, aberration effects, and the like, but these must necessarily be incidental to the development of the text itself. In *Atlas of Optical Phenomena* the roles are reversed: The pictures are the important things, and the text is secondary.[20]

d. Michelson, Albert A. *Studies in Optics.* (Reprint of 1927 Edition.) Chicago: University of Chicago Press, 1962. 176 pp.

In 1907 Michelson became the first American citizen to receive a Nobel Prize in science. His two great interests were the determination of the velocity of light, and interference effects.

e. Dogigli, Johannes. *The Magic of Rays.* New York: Alfred A. Knopf, 1961. 264 pp.

An interesting popularization of wave applications throughout the electromagnetic spectrum.

Although it champions the corpuscular theory of light, which at one time was generally discredited, the following is a classic in the field of optics:

Newton, Sir Isaac. *Opticks, or a Treatise of the Reflections, Refractions, Inflections and Colours of Light.* (Reprinted from the Fourth Edition, 1730.) London: G. Bell and Sons, 1931. 414 pp.

---

20. S. S. Ballard, in review. *Physics Today,* 16 (No. 9): 69, September 1963.

In the Introduction (pp. xi-xii), E. T. Whittaker states:

> The curious blending of corpuscular-theory with wave-theory which is suggested in some parts of his work, and which was a stumbling-block to the physicists of the nineteenth century, has been found to present considerable analogies with the modern views . . . So the volume which is here reprinted, after being esteemed for three generations chiefly as a historical landmark displaying a marvelous combination of theoretical and experimental skill, is now once more being read for its living scientific interest.

Three well-known textbooks on the theory of light are:

a. Houstoun, Robert A. *A Treatise on Light.* (Seventh Edition.) London: Longmans, Green and Company, 1938. 528 pp.
The format and illustrations are excellent in this well-designed textbook, which has been reprinted in 1943.

b. Preston, Thomas. *The Theory of Light.* (Fifth Edition.) London: The Macmillan Company, 1928. 643 pp.

c. Ditchburn, R. W. *Light.* (Second Edition.) New York: Interscience Publishers, 1963. 833 pp.

Among books which give considerable space to experimental aspects one finds:

a. Jenkins, Francis A., and White, Harvey E. *Fundamentals of Optics.* (Third Edition.) New York: McGraw-Hill Book Company, 1957. 637 pp.

b. Monk, George S. *Light; Principles and Experiments.* New York: McGraw-Hill Book Company, 1937. 477 pp.
In this intermediate text on all branches of optics are presented twenty-three experiments (pp. 343-416). A Dover Publications second edition (1963) added four appendices.

c. Valasek, Joseph. *Introduction to Theoretical and Experimental Optics.* New York: John Wiley and Sons, 1949. 454 pp.

Geometrical Optics

Rectilinear paths of light are the subject of:

a. Southall, James P. C. *Mirrors, Prisms and Lenses; A Textbook of Geometrical Optics.* (Third Edition.) New York: The Macmillan Company, 1933. 806 pp.

b. Martin, Louis C. *Geometrical Optics.* London: Sir Isaac Pitman and Sons, 1955. 215 pp.

c. Herzberger, Max. *Modern Geometrical Optics.* New York: Interscience Publishers, 1958. 504 pp.

d. Welford, W. T. *Geometrical Optics; Optical Instrumentation.* Amsterdam: North-Holland Publishing Company, 1962. 200 pp.
Geometrical optics is fundamental to design.

Lucid ray-diagrams also are featured in these smaller general texts:

a. Edser, Edwin. *Light for Students.* London: The Macmillan Company, 1931. 591 pp.

b. Bray, F. *Light.* (Second Edition.) London: Edward Arnold and Company, 1938. 369 pp.

Physical Optics

Phenomena associated with the vibrational wave characteristics of light are dealt with in physical optics, which has a comprehensive reference work of historic interest:

Gehrcke, Ernst. *Handbuch der Physikalischen Optik.* Leipzig: J. A. Barth, 1926-1928. 2 vols. in 5.

These books cover physical optics well:

a. Born, Max, and Wolf, Emil. *Principles of Optics: Electromagnetic Theory of Propagation, Interference and Diffraction of Light.* (Third Edition.) New York: Pergamon Press, 1965. 808 pp.

b. Rossi, Bruno B. *Optics.* Reading, Mass.: Addison-Wesley Publishing Company, 1957. 510 pp.

c. Robertson, John K. *Introduction to Optics, Geometrical and Physical.* (Fourth Edition.) New York: D. Van Nostrand Company, 1954. 416 pp.

d. Houstoun, Robert A. *Physical Optics.* New York: Interscience Publishers, 1958. 300 pp.

e. Stone, John M. *Radiation and Optics.* New York: McGraw-Hill Book Company, 1963. 544 pp.

f. Wood, Robert W. *Physical Optics.* (Third Edition.) New York: The Macmillan Company, 1934. 846 pp.

In stressing experimental aspects, the author attempts to present a "physical picture of the processes usually described by equations." Interferometers, refractometers, and other instruments are fully described.

For polarization, a standard presentation and an introduction:

a. Shurcliff, William A. *Polarized Light: Production and Use.* Cambridge, Mass.: Harvard University Press, 1962. 207 pp.

b. Shurcliff, William A., and Ballard, Stanley S. *Polarized Light.* Princeton, N. J.: D. Van Nostrand Company, 1964. 144 pp. (Momentum Book.)

See also "Resource Letter PL-1 on Polarized Light," by William A. Shurcliff. *American Journal of Physics,* 30: 227-230, March 1962.

The special topics of diffraction and scattering are treated in:

a. Meyer, Charles F. *The Diffraction of Light, X-rays, and Material Particles.* Chicago: University of Chicago Press, 1934. 473 pp.

(Also a "second revised edition": Ann Arbor, Mich.: J. W. Edwards, 1949. 473 pp.)

b. Van de Hulst, H. C. *Light Scattering by Small Particles.* New York: John Wiley and Sons, 1957. 470 pp.

For experiments and measurements in optics, see Chapter V.

See also index under Lasers.

Spectroscopy.

Early book and periodical material on spectroscopy may be found in the following indices:

Tuckerman, Alfred. *Index to the Literature of the Spectroscope.* Washington, D. C.: Smithsonian Institution, 1888. 423 pp. (Smithsonian Miscellaneous Collections, Vol. 32, No. 658.)

Index . . . 1887 through 1900. 1902. 373 pp. (Smithsonian Miscellaneous Collections, Vol. 41, No. 1312.)

A classic compilation of historic interest is:

Kayser, H. *Handbuch der Spectroscopie.* Leipzig: S. Hirzel, 1900-1934. 8 vols.

Vols. 5 and 6 are extensive collections of spectrographic data, revised in Vols. 7 and 8 for elements in the first part of the alphabet only. It is doubtful that publication will be completed.

Summaries may be found in:

Clark, George L., editor. *The Encyclopedia of Spectroscopy.* New York: Reinhold Publishing Corporation, 1960. 787 pp.

Useful tabulations are represented by the following:

a. Moore, C. E. *Atomic Energy Levels as Derived from the Analyses of Optical Spectra.*[21] Washington, D. C.: Government Printing Office, 1949-1958. 3 vols. (U. S. National Bureau of Standards Circular No. 467.)

Because the accumulation of data on atomic spectra has been so great since 1932, this set supersedes the Bacher and Goudsmit classic work, next cited.

b. Bacher, Robert F., and Goudsmit, Samuel. *Atomic Energy States, as Derived from the Analyses of Optical Spectra.* New York: McGraw-Hill Book Company, 1932. 562 pp.

c. Massachusetts Institute of Technology. *Wavelength Tables, with Intensities in Arc, Spark, or Discharge Tube of more than 100,000 Spectrum Lines Emitted by the Atomic Elements,* measured and com-

---

21. See also the related *Ultraviolet Multiplet Table* issued in 5 sections as NBS Circular No. 488.

piled under the direction of George R. Harrison. New York: John Wiley and Sons, 1939. 429 pp.

d. Zaidel', A. N.; Prokof'ev, V. K.; and Raiskii, S. M. *Tables of Spectrum Lines.* Berlin: VEB Verlag Technik; New York: Pergamon Press, 1961. 550 pp.

e. *Tables of Spectral-Line Intensities,* compiled by William F. Meggers, Charles H. Corliss and Bourdon F. Scribner. Washington, D. C.: Government Printing Office, 1961-1962. 2 vols. (U. S. National Bureau of Standards Monograph 32.)
Volume arrangement is by Elements and Wavelengths, respectively.

References to additional tables may be found in Sawyer's *Experimental Spectroscopy;* also in Brode's *Chemical Spectroscopy,*[22] and in the Herzberg series, below.

Well-documented textbooks prepared with the beginner's needs in view are:

a. Herzberg, Gerhard. *Atomic Spectra and Atomic Structure.*[23] New York: Prentice-Hall, Inc., 1937. 257 pp.

b. Herzberg, Gerhard. *Molecular Spectra and Molecular Structure.* Princeton, N. J.: D. Van Nostrand Company, 1945-1966. 3 vols., as follows:

Vol. 1: *Spectra of Diatomic Molecules.* (Second Edition.) 1950. 658 pp.

Vol. 2: *Infrared and Raman Spectra of Polyatomic Molecules.* 1945. 632 pp.

Vol. 3: *Electronic Spectra and Electronic Structure of Polyatomic Molecules.* 1966. 875 pp.

c. Wilson, E. Bright, Jr.; Decius, J. C.; and Cross, P. C. *Molecular Vibrations; The Theory of Infrared and Raman Vibrational Spectra.* New York: McGraw-Hill Book Company, 1955. 388 pp.

d. Houghton, J. T., and Smith, S. D. *Infra-Red Physics.* New York: Oxford University Press, 1966. 319 pp.
Spectroscopy and physical phenomena.

e. Brügel, Werner. *An Introduction to Infrared Spectroscopy.* London: Methuen and Company, 1962. 419 pp.

f. Sommerfeld, Arnold. *Atomic Structure and Spectral Lines,* translated from fifth German edition. (Third Edition.) London: Methuen

---

22. W. R. Brode, *Chemical Spectroscopy,* pp. 74-84. (Second Edition.) New York: John Wiley and Sons, 1943.

23. A reprint issued by Dover Publications (New York, 1944) incorporates a few changes and additions.

and Company, 1934; Supp.: *Wave Mechanics.*[24] London: Methuen and Company, 1930.

g. Kuhn, H. G. *Atomic Spectra.* New York: Academic Press, 1962. 436 pp.

Physicist's approach, stressing correspondence between classical and quantum physics.

h. Whiffen, D. H. *Spectroscopy.* New York: John Wiley and Sons, 1966. 205 pp.

More advanced, and still valuable for its fundamental theory, tables, etc., is the classic treatise:

Condon, E. U., and Shortley, G. H. *The Theory of Atomic Spectra.* Cambridge, England: At the University Press, 1951. 441 pp. (Reprinted with corrections.)

Techniques employing electrical waves in the 0.5 mm. to 10 cm. range are described in:

a. Townes, C. H., and Schawlow, A. L. *Microwave Spectroscopy.* New York: McGraw-Hill Book Company, 1955. 698 pp.

b. Gordy, Walter; Smith, William V.; and Trambarulo, Ralph F. *Microwave Spectroscopy.* New York: John Wiley and Sons, 1953. 446 pp.

Other techniques use high-speed electron streams, and penetrating waves extending higher than X-rays in frequency:

Siegbahn, Kai, editor. *Alpha- Beta- and Gamma-Ray Spectroscopy.* Amsterdam: North-Holland Publishing Company, 1965. 2 vols.

This is an expanded edition (1742 pp.) of his *Beta- and Gamma-Ray Spectroscopy* published a decade ago.

See also index entries under Spectroscopy.

Physiological optics.

Paralleling similar work by the same author in the field of sound, we have the classic treatise on the physical, physiological, and psychological effects of light:

Helmholtz, Hermann L. F. von. *Treatise on Physiological Optics,* translated from the 3d German edition. Edited by James P. C. Southall. New York: Optical Society of America, 1924-1925. 3 vols.

Two presentations by physicists are:

a. Southall, James P. C. *Introduction to Physiological Optics.* London: Oxford University Press, 1937. 426 pp.

---

24. See fourth German edition (Braunschweig: Friedrich Vieweg und Sohn, 1960).

b. Houstoun, Robert A. *Vision and Colour Vision.* London: Longmans, Green and Company, 1932. 238 pp.

From physiological standpoints, we have:

a. Bartley, S. H. *Vision; A Study of the Basis.* New York: D. Van Nostrand Company, 1941. 350 pp.

b. Luckiesh, Matthew,[25] and Moss, Frank K. *The Science of Seeing.* New York: D. Van Nostrand Company, 1937. 548 pp.

c. Graham, Clarence H., *et al. Vision and Visual Perception.* New York: John Wiley and Sons, 1965. 637 pp.

d. Davson, Hugh, editor. *The Eye.* New York: Academic Press, 1962. 4 vols.

Color.

Books on colorimetry, the measurement of color, have already been listed under Special Measurements in Chapter V. Reference was made to the unreliability of color specification by color samples. However, collections of color chips have practical uses for purposes of matching, combination and arrangement, and are found in:

a. Jacobson, Egbert; Granville, Walter C.; and Foss, Carl E. *Color Harmony Manual.* (Third Edition.) Chicago: Container Corporation of America, 1948. 51 pp. text and 37 charts containing 973 separate color samples. Color designations of the detachable hexagon chips differ from the previous editions. (See book review in *Physics Today,* 3 (No. 8): 34-36, August 1950.)

b. Jacobson, Egbert. *Basic Color; An Interpretation of the Ostwald Color System.* Chicago: Paul Theobald, 1948. 207 pp.

c. Munsell, A. H. *Munsell Book of Color.* (Standard Edition.) Baltimore: Munsell Color Company, 1929. 2 vols.
This is a superb oversize set with fine color plates in the second volume.

d. Maerz, Aloys, and Paul, M. R. *A Dictionary of Color.* (Second Edition.) New York: McGraw-Hill Book Company, 1950. 206 pp. This book of plates serves to correlate colors with their common names. The pages are marked off in squares, carefully color-graded, with names printed on dummy sheets opposite. Bibliographies, definitions, and techniques are included.

Terms used in English industry and science are coordinated and compared with American usage in:

Physical Society (London). *Report on Colour Terminology.* London: The Society, 1948. 56 pp.

---

25. See also his *Reading as a Visual Task.*

An American committee report furnishes an authoritative and comprehensive reference work on all phases of color:

Optical Society of America. Committee on Colorimetry. *The Science of Color.* New York: Thomas Y. Crowell Company, 1953. 385 pp.

There is a corrected 5th printing (Edwards Bros., 1966). Physical, psychological and physiological aspects are treated, in addition to colorimetry *per se.*

Interesting introductory treatments of color theory and practice are:

a. Evans, Ralph M. *An Introduction to Color.* New York: John Wiley and Sons, 1948. 340 pp.

The author's purpose is stated (p. v):

> Color sprawls across the three enormous subjects of physics, physiology, and psychology. In the past it has been rare that any intensive worker in color has had the opportunity of understanding all three phases. It is to fill this gap that the book has been written.

b. Judd, Deane B., and Wyszecki, Günter. *Color in Business, Science, and Industry.* (Second Edition.) New York: John Wiley and Sons, 1963. 500 pp.

c. Wyszecki, Günter, and Stiles, W. S. *Color Science: Concepts and Methods, Quantitative Data and Formulas.* New York: John Wiley and Sons, 1967. 628 pp.

d. Houstoun, Robert A. *Light and Colour.* London: Longmans, Green and Company, 1923. 179 pp.

e. Burnham, R. W.; Hanes, R. M.; and Bartleson, C. James. *Color: A Guide to Basic Facts and Concepts.* New York: John Wiley and Sons, 1963. 249 pp.

f. Murray, Humphrey D., editor. *Colour in Theory and Practice.* (New Edition.) London: Chapman and Hall, Ltd., 1952. 360 pp.

For history of color, see Halbertsma's book, and Birren's.

Luminescence embraces the color phenomena of fluorescence and phosphorescence. Some substances exposed to short waves (like ultraviolet or X-rays) absorb them and emit longer waves. If such fluorescence persists after removal of the light source, it is termed phosphorescence. Useful references follow:

a. Pringsheim, Peter. *Fluorescence and Phosphorescence.* New York: Interscience Publishers, 1949. 794 pp.

b. Leverenz, Humboldt W. *An Introduction to Luminescence of Solids.* New York: John Wiley and Sons, 1950. 569 pp.

c. American Physical Society. *Properties and Characteristics of*

Solid Luminescent Materials. New York: John Wiley and Sons, 1948. 459 pp.

d. Goldberg, Paul, editor. *Luminescence of Inorganic Solids.* New York: Academic Press, 1966. 765 pp.

e. Thornton, P. R. *The Physics of Electroluminescent Devices.* London: E. and F. N. Spon, 1967. 382 pp.

f. Passwater, Richard A. *Guide to Fluorescence Literature.* New York: Plenum Press, 1967. 367 pp.
Luminescence is covered for the period 1950-1964.

(Luminescent materials had been included in *Semiconductor Abstracts,* compiled by the Battelle Memorial Institute for the years 1953-1959, and published by John Wiley and Sons, 1955-1962.)

Applied optics.

Technical presentations that deal more thoroughly with optical instrument design than the descriptive books cited in Chapter V under Instruments include:

a. Kingslake, Rudolf, editor. *Applied Optics and Optical Engineering.* New York: Academic Press, 1965- 5 vols., in process.
Volume titles are: 1, Light; its generation and modification; 2, The detection of light and infrared radiation; 3, Optical components; and 4-5, Optical instruments.

b. Martin, Louis C. *Technical Optics.* (Revised and Enlarged Edition.) [26] London: Sir Isaac Pitman and Sons, 1948-1950. 2 vols.

c. Conrady, A. E. *Applied Optics and Optical Design.* Part One. London: Oxford University Press, 1929. 518 pp.; Part Two, edited and completed by Rudolf Kingslake. New York: Dover Publications, 1960. 323 pp.

d. Czapski, Siegfried, and Eppenstein, Otto. *Grundzüge der Theorie der Optischen Instrumente nach Abbe.*[27] (Dritte Auflage.) Leipzig: J. A. Barth, 1924. 747 pp.
This well-illustrated series of articles embodies the work of Abbe and others associated with the famous Zeiss works.[28]

---

26. Of his *An Introduction to Applied Optics.* London: Sir Isaac Pitman and Sons, 1930-1932. 2 vols.
27. See also M. v. Rohr, *Geometrical Investigation of the Formation of Images in Optical Instruments.* (Vol. 1 of "The Theory of Optical Instruments.") London: H. M. Stationery Office, 1920. 612 pp.
28. See F. Auerbach, *The Zeiss Works and the Carl Zeiss Foundation in Jena.* (Translated from the fifth German edition.) London: W. and G. Foyle, Ltd., 1927. 273 pp.

e. Smith, Warren J. *Modern Optical Engineering: The Design of Optical Systems.* New York: McGraw-Hill Book Company, 1966. 468 pp.
Scientists may apply this to their own research.

f. Bracey, R. J. *The Technique of Optical Instrument Design.* London: The English Universities Press, 1960. 316 pp.

Illuminating engineering is beyond the scope of this guide, but a few useful references follow:

a. Moon, Parry H., and Spencer, Domina E. *Lighting Design.* Cambridge, Mass.: Addison-Wesley Press, 1948. 482 pp.

b. Kraehenbuehl, John O. *Electric Illumination.* (Second Edition.) New York: John Wiley and Sons, 1951. 446 pp.

c. Barrows, William E. *Light, Photometry and Illuminating Engineering.* (Third Edition.) New York: McGraw-Hill Book Company, 1951. 415 pp.

d. Phillips, Derek. *Lighting in Architectural Design.* New York: McGraw-Hill Book Company, 1964. 312 pp.

Ultraviolet effects and applications may be found in:

a. Koller, Lewis R. *Ultraviolet Radiation.* (Second Edition.) New York: John Wiley and Sons, 1965. 312 pp.

b. Green, A. E. S., editor. *The Middle Ultraviolet: Its Science and Technology.* New York: John Wiley and Sons, 1966. 390 pp.
Solar, atmospheric, atomic and molecular aspects are presented.

Infrared technology has gone beyond "classical" infrared spectroscopy treated previously:

a. Jamieson, John A., *et al.* *Infrared Physics and Engineering.* New York: McGraw-Hill Book Company, 1963. 673 pp.

b. Kruse, Paul W., *et al.* *Elements of Infrared Technology: Generation, Transmission and Detection.* New York: John Wiley and Sons, 1962. 448 pp.

c. Hackforth, Henry L. *Infrared Radiation.* New York: McGraw-Hill Book Company, 1960. 303 pp.

Physical effects of light are discussed in:

Seliger, Howard H., and McElroy, William D. *Light: Physical and Biological Action.* New York: Academic Press, 1965. 417 pp.

Chemical effects and physical optics in photography are discussed in:

a. Dhar, Nil R. *The Chemical Action of Light.* London: Blackie and Son, 1931. 512 pp.

b. James, T. H., editor. *The Theory of the Photographic Process.* (Third Edition.) New York: The Macmillan Company, 1966. 591 pp.

One may supplement this comprehensive survey by reference to the earlier editions by photographic pioneer C. E. K. Mees, who stated:

> The theory of the photographic process involves a study of the light-sensitive layers used, of the factors which control their sensitivity to light, of the changes induced in them by the action of light, of the nature of development, and of the properties of the final image and its relation in tone values to the tone values of the scene from which it was produced.

c. Mack, Julian E., and Martin, Miles J. *The Photographic Process.* New York: McGraw-Hill Book Company, 1939. 586 pp.
This is a college text from the scientific viewpoint, and includes thirty-one experiments (pp. 523-568), e.g., Perspective and angle of view; Exposure; Photomicrography.

d. Boucher, Paul E. *Fundamentals of Photography.* (Fourth Edition.) Princeton, N. J.: D. Van Nostrand Company, 1963. 535 pp.
The physics and chemistry of photography are stressed. There are laboratory experiments on lenses and diaphragms, aberrations, infrared photography, etc.

e. Franke, Georg. *Physical Optics in Photography.* London: Focal Press, Ltd., 1966. 218 pp.
Design of optical systems.

See also C. W. Miller, "Photography in the Physics Curriculum." *American Journal of Physics,* 9: 107-110, April 1941; and Eastman Kodak's *Abstracts of Photographic Science and Engineering Literature,* formerly *Monthly Abstract Bulletin.*

Of related photographic interest are "Pictorial devices" (see subject index), and books by M. Cagnet, W. Gentner, C.-N. Martin, C. F. Powell, and G. D. Rochester (via author index).

6—Electricity and Magnetism

General.

There is available a treatise of historic interest:

Graetz, L., editor. *Handbuch der Elektrizität und des Magnetismus.* Leipzig: J. A. Barth, 1918-1928. 5 vols.
A large staff of specialists has contributed signed articles, with copious bibliographical citation. Volume titles are: (Vol. 1) Elektrizitätserregung und elektrostatik; (Vol. 2) Stationäre strome; (Vol. 3) Elektronen und ionen; (Vol. 4) Magnetismus und elektromagnetismus; and (Vol. 5) Zeitliche vorgänge. Technik.

Bibliographies include a historical compilation by a master bibliographer:

Mottelay, Paul F. *Bibliographical History of Electricity and Magnetism.* London: Charles Griffin and Company, 1922. 673 pp.
It is chronologically arranged, from 2637 B.C. to 1821 A.D., and incorporates biographical, historical and bibliographical material serving to identify early works.

Books appearing during an eventful quarter-century are listed in:

May, G. *A Bibliography of Electricity and Magnetism 1860 to 1883 with Special Reference to Electro-technics.* London: Trübner and Company, 1884. 203 pp. (Also a volume covering 1876-1885, published in London by Whitaker in 1886.)

Textbooks on electricity and magnetism are so numerous as to make selection of the following typical ones difficult:

a. Pugh, Emerson M., and Pugh, Emerson W. *Principles of Electricity and Magnetism.* Reading, Mass.: Addison-Wesley Publishing Company, 1960. 430 pp.

b. Scott, William T. *The Physics of Electricity and Magnetism.* (Second Edition.) New York: John Wiley and Sons, 1966. 703 pp.

c. Benumof, Reuben. *Concepts in Electricity and Magnetism.* New York: Holt, Rinehart and Winston, 1961. 374 pp.

d. Page, Leigh, and Adams, Norman I. *Principles of Electricity; An Intermediate Text in Electricity and Magnetism.* (Third Edition.) Princeton, N. J.: D. Van Nostrand Company, 1958. 533 pp.
Principles rather than applications are stressed.

e. Harnwell, Gaylord P. *Principles of Electricity and Electromagnetism.* (Second Edition.) New York: McGraw-Hill Book Company, 1949. 670 pp.
Classical phenomena and modern developments are outlined, with emphasis upon experimental aspects.

f. Winch, Ralph P. *Electricity and Magnetism.* (Second Edition.) Englewood Cliffs, N. J.: Prentice-Hall, Inc., 1963. 606 pp.
Physical aspects are kept in the foreground.

For experiments and measurements in electricity, see Chapter V.

Electromagnetic theory.

Modern field theory stems largely from the pioneer work embodied in the following classic treatise:

Maxwell, James Clerk. *A Treatise on Electricity and Magnetism.* (Third Edition.) Oxford: At the Clarendon Press, 1904. 2 vols. (Also available in a thousand-page volume reissued by Dover Publications, New York, in 1954.)

See also his *Scientific Papers* (Cambridge University Press, 1890) very compactly reproduced in one volume by Dover Publications in 1952.

Those who wish an interesting glimpse of the formative stages of Maxwell's contributions may refer to:

Larmor, Sir Joseph. *Origins of Clerk Maxwell's Electric Ideas as Described in Familiar Letters to William Thomson.* Cambridge, England: At the University Press, 1937. 56 pp.

"The letters now published present a sharp and crisp account of the genesis and rapid progress of Clerk Maxwell's ideas as he groped towards a structural theory of the electric and magnetic field.

The theory was expanded and clarified by Heaviside,[29] who further developed the mathematical methods involved:

Heaviside, Oliver. *Electromagnetic Theory.* (Complete and unabridged edition of Volumes I, II, III.) New York: Dover Publications, 1950. 386 pp.

This reproduction of the classic treatise (first published 1893-1912) has a biography by Ernst Weber.

Widely used expositions of the subject are:

a. Jeans, Sir James H. *The Mathematical Theory of Electricity and Magnetism.* (Fifth Edition.) Cambridge, England: At the University Press, 1941. 652 pp.

b. Sauter, Fritz, editor. *Electromagnetic Fields and Interactions.* New York: Blaisdell Publishing Company, 1964. 2 Vols., as follows:

Vol. 1: *Electromagnetic Theory and Relativity,* by Richard Becker. 439 pp.

Vol. 2: *Quantum Theory of Atoms and Radiation,* by Richard Becker. 403 pp.

The first volume resembles the English edition of *The Classical Theory of Electricity and Magnetism,* by Max Abraham and Richard Becker. Both stem from many German editions. A new third volume is contemplated by the editor.

c. Panofsky, Wolfgang K. H., and Phillips, Melba. *Classical Electricity and Magnetism.* (Second Edition.) Reading, Mass.: Addison-Wesley Publishing Company, 1962. 494 pp.

d. Stratton, Julius A. *Electromagnetic Theory.* New York: McGraw-Hill Book Company, 1941. 615 pp.

e. Smythe, William R. *Static and Dynamic Electricity.* (Second Edition.) New York: McGraw-Hill Book Company, 1950. 616 pp.

---

29. For other Heaviside contributions, see L. Cohen's and E. J. Berg's books.

Historical considerations are retraced in:

O'Rahilly, Alfred. *Electromagnetics; A Discussion of Fundamentals.* London: Longmans, Green and Company, 1938. 884 pp. An excellent selective bibliography on pp. 860-878 may be used to supplement Mottelay's work. This is a critical treatise by a mathematical physicist.

A concise but clear résumé is furnished by:

Duff, A. Wilmer, and Plimpton, Samuel J. *Elements of Electro-Magnetic Theory.* Philadelphia: The Blakiston Company, 1940. 173 pp.

More comprehensive textbooks include:

a. Whitmer, Robert M. *Electromagnetics.* (Second Edition.) Englewood Cliffs, N. J.: Prentice-Hall, Inc., 1962. 357 pp.

b. Reitz, John R., and Milford, Frederick J. *Foundations of Electromagnetic Theory.* (Second Edition.) Reading, Mass.: Addison-Wesley Publishing Company, 1967. 435 pp.

c. Slater, John C., and Frank, Nathaniel H. *Electromagnetism.* New York: McGraw-Hill Book Company, 1947. 240 pp.

d. Barut, A. O. *Electrodynamics and Classical Theory of Fields and Particles.* New York: The Macmillan Company, 1964. 230 pp.

e. Schelkunoff, Sergei A. *Electromagnetic Fields.* New York: Blaisdell Publishing Company, 1963. 413 pp.

f. Jackson, John D. *Classical Electrodynamics.* New York: John Wiley and Sons, 1962. 641 pp.

See also "Resource Letter FC-1 on the Evolution of the Electromagnetic Field Concept," by William T. Scott. *American Journal of Physics,* 31: 819-826, November 1963.

Electromagnetic waves [30] that make radio communication possible, are discussed in:

a. Collin, Robert E. *Field Theory of Guided Waves.* New York: McGraw-Hill Book Company, 1960. 606 pp.

b. Chodorow, Marvin, and Susskind, Charles. *Fundamentals of Microwave Electronics.* New York: McGraw-Hill Book Company, 1964. 297 pp.

c. Skilling, Hugh H. *Fundamentals of Electric Waves.* (Second Edition.) New York: John Wiley and Sons, 1948. 245 pp.

d. Corson, Dale R., and Lorrain, Paul. *Introduction to Electro-*

---

30. For a unified mathematical treatment of the common types of wave motion, see C. A. Coulson, *Waves.* (Seventh Edition.) New York: Interscience Publishers, 1955. 171 pp.

magnetic Fields and Waves. San Francisco: W. H. Freeman and Company, 1962. 552 pp.

e. Ramo, Simon; Whinnery, John R.; and Van Duzer, Theodore. *Fields and Waves in Communication Electronics.* New York: John Wiley and Sons, 1965. 754 pp.

See also books on radio mentioned in Chapter X—Electronics.

Circuits.

Heaviside's work again comes to view in:

Cohen, Louis. *Heaviside's Electrical Circuit Theory.* New York: McGraw-Hill Book Company, 1928. 169 pp.

Cohen states the purpose of his presentation (p. v) as follows:

> The importance of Heaviside's contributions to electrical theory is now generally recognized and appreciated. His teachings nevertheless are available to only a comparatively few; to the many engineers and physicists who could profit much by it, the work of Heaviside is more or less a sealed book.

Practical considerations place electric circuits within the purview of electrical engineering textbooks, such as:

a. Skilling, Hugh H. *Electrical Engineering Circuits.* (Second Edition.) New York: John Wiley and Sons, 1965. 783 pp.

b. Romanowitz, H. Alex. *Electrical Fundamentals and Circuit Analysis.* New York: John Wiley and Sons, 1966. 715 pp.

c. Brenner, Egon, and Javid, Mansour. *Analysis of Electric Circuits.* (Second Edition.) New York: McGraw-Hill Book Company, 1967. 724 pp.

d. Siskind, Charles S. *Electrical Circuits.* (Second Edition.) New York: McGraw-Hill Book Company, 1964. 600 pp.

e. Pearson, S. Ivar, and Maler, George J. *Introductory Circuit Analysis.* New York: John Wiley and Sons, 1965. 546 pp.

See also Electronic Circuits.

Properties.

Magnetic, dielectric, thermoelectric and piezoelectric phenomena are associated with certain kinds of materials. For magnetism, the following group of treatises covers varied aspects:

a. Stoner, Edmund C. *Magnetism and Matter.* London: Methuen and Company, 1934. 575 pp.

Magnetic properties of matter in general are discussed.

b. Williams, Samuel R. *Magnetic Phenomena; An Elementary Treatise.* New York: McGraw-Hill Book Company, 1931. 230 pp.

The author seeks to encourage research in such areas as Magneto-magnetics, Magneto-mechanics, Magneto-acoustics, Magneto-optics, etc.

c. Williams, D. E. G. *The Magnetic Properties of Matter.* New York: American Elsevier Publishing Company, 1966. 232 pp.

d. Rado, George T., and Suhl, Harry, editors. *Magnetism.* New York: Academic Press, 1963-1966. 4 vols.
Fifty authorities produced this encyclopedic set.

e. Morrish, Allan H. *The Physical Principles of Magnetism.* New York: John Wiley and Sons, 1965. 680 pp.

f. Bates, Leslie F. *Modern Magnetism.* (Fourth Edition.) Cambridge, England: At the University Press, 1961. 514 pp.
Experimental aspects are stressed for greater clarity of presentation.

Magnetism and Magnetic Materials conferences sponsored by the American Institute of Physics and electrical engineering groups have published proceedings, materials digests, and literature indexes.

For low-temperature magnetic effects see books by Casimir and Garrett.

For dielectrics (insulating materials in condensers) one may consult:

a. Fröhlich, Herbert. *Theory of Dielectrics.* (Second Edition.) London: Oxford University Press, 1958. 192 pp.

b. Von Hippel, Arthur R. *Dielectrics and Waves.* New York: John Wiley and Sons, 1954. 284 pp.

c. O'Dwyer, J. J. *The Theory of Dielectric Breakdown of Solids.* Oxford: At the Clarendon Press, 1964. 142 pp.

d. Kok, J. A. *Electrical Breakdown of Insulating Liquids.* New York: Interscience Publishers, 1961. 132 pp.

e. Whitehead, S. *Dielectric Phenomena.* London: Ernest Benn, Ltd., 1927-1932. 3 vols.
The coverage by volumes is: (1) Electrical discharges in gases; (2) Electrical discharges in liquids; and (3) Breakdown of solid dielectrics.

f. National Research Council. Conference on Electrical Insulation. *Digest of Literature on Dielectrics,* 1936 to date. Washington, D. C.: The Council, 1937 to date.

Thermoelectricity is covered by:

a. MacDonald, D. K. C. *Thermoelectricity; An Introduction to the Principles.* New York: John Wiley and Sons, 1962. 133 pp.

b. Egli, Paul H., editor. *Thermoelectricity.* New York: John Wiley and Sons, 1960. 407 pp.

c. Harman, T. C., and Honig, J. M. *Thermoelectric and Thermomagnetic Effects and Applications.* New York: McGraw-Hill Book Company, 1967. 377 pp.

See also index under Thermionics.

For piezoelectricity (electrical effects produced by pressure) we have:

a. Cady, Walter G. *Piezoelectricity; An Introduction to the Theory and Applications of Electromechanical Phenomena in Crystals.*[31] New York: McGraw-Hill Book Company, 1946. 806 pp.
Historical development is sketched.

b. Mason, Warren P. *Piezoelectric Crystals and their Application to Ultrasonics.* New York: D. Van Nostrand Company, 1950. 508 pp.

Applied electricity.

Electrical engineering is beyond this *Guide's* scope, but it is nevertheless based on physics, as Timbie and Bush agree:

> Fundamental physics which forms the basis for electrical engineering has made striking advances. Some of these, particularly in the field of atomistics, have been spectacular. A far better understanding is being secured of the difficult phenomena of the conduction of electricity through gases, liquids, and solids. The physics basis of electrical engineering is becoming more extended and better established.[32]

This relationship gives title to:

Martin, Thomas L., Jr. *Physical Basis for Electrical Engineering.* Englewood Cliffs, N. J.: Prentice-Hall, Inc., 1957. 410 pp.

As further sources, one may consult the electrical engineering handbooks by Knowlton and Pender, *et al.,* as well as textbooks like the following:

a. Del Toro, Vincent. *Principles of Electrical Engineering.* Englewood Cliffs, N. J.: Prentice-Hall, Inc., 1965. 775 pp.

b. Gray, Alexander, and Wallace, G. A. *Principles and Practice of Electrical Engineering.* (Eighth Edition.) New York: McGraw-Hill Book Company, 1962. 616 pp.

c. Rosenblatt, Jack, and Friedman, M. H. *Direct and Alternating Current Machinery.* New York: McGraw-Hill Book Company, 1963. 432 pp.

---

31. A 2-vol. (822 pp.) Dover Publications reprint (1964) has an appendix on recent progress.
32. W. H. Timbie and V. Bush, *Principles of Electrical Engineering,* p. v. (Fourth Edition.) New York: John Wiley and Sons, 1951.

7—Electronics

General.

The present discussion centers upon electron flow rather than electrons as units of atomic structure, which are treated in Chapter X—Molecular . . . Physics. As previously noted, this field has its own *Electronic Engineering Master Index* in the form of multi-year volumes.

Further bibliographical aid is provided by:

a. Nottingham, Wayne B., *et al. Bibliography on Physical Electronics.* Cambridge, Mass.: Research Laboratory of Electronics, Massachusetts Institute of Technology, 1954. 428 pp. (Addison-Wesley Publishing Company, distributor.)
This is a classified listing of books and articles on free electron phenomena, for period 1930-1950. *A Bibliographical Guide.* London: Macdonald and Company, 1961-1965. 2 vols.
Coverage is 1945-1964.
Over three thousand items are included, 1945-1959.

c. *Electronic Properties of Materials: A Guide to the Literature,* edited by H. T. Johnson (Vol. 1) and D. L. Grigsby (Vol. 2). New York: Plenum Press, 1965-1967. 2 vols. in 4.
From materials and properties in each Part 1, numbered reference is made to Part 2 bibliographies.

There is a useful encyclopedia:

Susskind, Charles, editor. *Encyclopedia of Electronics.* New York: Reinhold Publishing Corporation, 1962. 974 pp.

Well-rounded introductions are furnished by:

a. Malmstadt, H. V.; Enke, C. G.; and Toren, E. C., Jr. *Electronics for Scientists; Principles and Experiments for Those Who Use Instruments.* New York: W. A. Benjamin, Inc., 1962. 619 pp.

b. Hunten, Donald M. *Introduction to Electronics.* New York: Holt, Rinehart and Winston, 1964. 369 pp.

Additional general books selected from many:

a. Alfrey, G. F. *Physical Electronics.* Princeton, N. J.: D. Van Nostrand Company, 1964. 220 pp.

b. Hemenway, Curtis L.; Henry, Richard W.; and Caulton, Martin. *Physical Electronics.* (Second Edition.) New York: John Wiley and Sons, 1967. 449 pp.

c. Owen, George E., and Keaton, P. W. *Fundamentals of Electronics.* New York: Harper and Row, 1966-1967. 3 vols.

ERRATUM

p. 208:

Nottingham entry near top of page should end with the words "phenomena, for period 1930-1950."

Then start new item as follows:

b. Moore, C. K., and Spencer, K. J. *Electronics: A Bibliographical Guide*. London: Macdonald and Company, 1961-1965. 2 vols. Coverage is 1945-1964; further volumes contemplated.

A similar selection for solid-state electronics:

a. Beam, Walter R. *Electronics of Solids.* New York: McGraw-Hill Book Company, 1965. 633 pp.

b. Wang, S. *Solid State Electronics.* New York: McGraw-Hill Book Company, 1966. 778 pp.

c. Azároff, Leonid V., and Brophy, James J. *Electronic Processes in Materials.* New York: McGraw-Hill Book Company, 1963. 462 pp.

d. Stringer, John. *An Introduction to the Electron Theory of Solids.* Oxford, New York, etc.: Pergamon Press, 1967. 246 pp.

See also index under Solid state.

For experiments and measurements in electronics, see Chapter V.

Conduction in gases.

Ionization occurs when outer electrons are dislodged by collision or radiation and become current carriers. Classical works in this area are:

a. Townsend, Sir John S. E. *Electrons in Gases.* London: Hutchinson's Scientific and Technical Publications, 1947. 166 pp.
This carries forward work described in his earlier *Electricity in Gases* (1915).

b. Thomson, Sir Joseph J., and Thomson, Sir George P. *Conduction of Electricity through Gases.* (Third Edition.) Cambridge, England: At the University Press, 1928-1933. 2 vols.
The first edition, appearing in 1903, had named this field of experimentation.

More recent experimental research is described by an outstanding specialist in the field:

a. Loeb, Leonard B. *Basic Processes of Gaseous Electronics.* (Second Edition.) Berkeley, Cal.: University of California Press, 1961. 1028 pp.

b. Loeb, Leonard B. *Electrical Coronas; Their Basic Physical Mechanisms.* Berkeley, Cal.: University of California Press, 1965. 694 pp.

Other noteworthy textbooks are:

a. Francis, Gordon. *Ionization Phenomena in Gases.* London: Butterworths Scientific Publications, 1960. 300 pp.

b. Brown, Sanborn C. *Introduction to Electrical Discharges in Gases.* New York: John Wiley and Sons, 1966. 286 pp.

c. Engel, A. von, and Steenbeck, M. *Elektrische Gasentladungen; Ihre Physik und Technik.* Berlin: Springer, 1932-1934. 2 vols.
This is characterized as "admirable" by Loeb. For an introductory treatment see von Engel's *Ionized Gases.* (Second Edition.) Oxford: At the Clarendon Press, 1965. 325 pp.

d. Meek, J. M., and Craggs, J. D. *Electrical Breakdown of Gases.* London: Oxford University Press, 1954. 507 pp.

e. Llewellyn-Jones, F. *Ionization and Breakdown in Gases.* New York: John Wiley and Sons, 1957. 176 pp.

Gas discharge tables are found in:

Knoll, M., Ollendorff, F.; and Rompe, R. *Gasentladungstabellen; Tabellen, Formeln und Kurven zur Physik und Technik der Elektronen und Ionen.* Berlin: Springer, 1935. 171 pp.

See also index under Collision phenomena.

Plasma physics.

In a plasma, many electrons and ions are not bound into neutral atoms. The development of plasma research has been phenomenal and multifarious:

> Plasma physics is the protean and fashionable discipline concerned with the behavior of systems composed mainly of electrons and ions in which the electromagnetic forces between the charged particles and the effect of electromagnetic fields and boundaries that may be present are important in determining the behavior of the systems. Since many people now use the term "plasma physics" synonymously with such venerable branches of physics as surface-gas interactions, gaseous discharges, arc discharges, glow-discharge phenomena, and electron-beam dynamics, and with such novel fields as controlled thermonuclear fusion, ion and plasma propulsion, cosmic and ionospheric physics, and magnetohydrodynamics, the use of the adjective "protean" to characterize plasma physics is fully justified.[33]

Superb introductory expositions featuring physical insights are:

a. Spitzer, Lyman, Jr. *Physics of Fully Ionized Gases.* (Second Edition.) New York: Interscience Publishers, 1962. 170 pp.

b. Rose, David J., and Clark, Melville. *Plasmas and Controlled Fusion.* New York: John Wiley and Sons, 1961. 493 pp.

Useful general presentations include:

a. Holt, E. H., and Haskell, R. E. *Foundations of Plasma Dynamics.* New York: The Macmillan Company, 1965. 510 pp.

b. Longmire, Conrad L. *Elementary Plasma Physics.* New York: Interscience Publishers, 1963. 296 pp.
Approximation methods still basic to plasma physics are described and evaluated.

c. Uman, Martin A. *Introduction to Plasma Physics.* New York: McGraw-Hill Book Company, 1964. 226 pp.

---

33. H. Chang, in book review. *Physics Today,* 15 (No. 6): 58, June 1962.

d. Kunkel, Wulf B., editor. *Plasma Physics in Theory and Application.* New York: McGraw-Hill Book Company, 1966. 469 pp.

For wave phenomena:

a. Stix, Thomas H. *The Theory of Plasma Waves.* New York: McGraw-Hill Book Company, 1962. 283 pp.

b. Brandstatter, Julius J. *An Introduction to Waves, Rays, and Radiation in Plasma Media.* New York: McGraw-Hill Book Company, 1963. 704 pp.

See also the preceding section on Conduction in gases; index under Magnetohydrodynamics; and Gray's survey.[34]

See also "Resource Letter PP-1 on Plasma Physics," by Sanborn C. Brown. *American Journal of Physics,* 30: 303-307, April 1962.

See also "Resource Letter PP-2 on Plasma Physics: Waves and Radiation Processes in Plasmas," by G. Bekefi and Sanborn C. Brown. *American Journal of Physics,* 34: 1001-1005, November 1966.

Electron optics.

The behavior of electron beams in magnetic fields is described in:

a. Cosslett, Vernon E. *Introduction to Electron Optics.* (Second Edition.) Oxford: At the Clarendon Press, 1950. 293 pp.

b. Klemperer, Otto. *Electron Optics.* (Second Edition.) Cambridge, England: At the University Press, 1953. 471 pp.

c. Sturrock, P. A. *Static and Dynamic Electron Optics.* Cambridge, England: At the University Press, 1955. 240 pp.

d. Glaser, Walter. *Grundlagen der Elektronenoptik.* Vienna: Springer, 1952. 699 pp.

e. Grivet, P. *Electron Optics.* New York: Pergamon Press, 1965. 781 pp.

See also index under Beams, particle.

For an important application to scientific research, electron microscopy, one may consult the following bibliographies and surveys:

a. Cosslett, Vernon E. *Bibliography of Electron Microscopy.* London: Longmans, Green and Company, 1951. 350 pp.

This includes 2500 article references through 1948, arranged by author, with contents notes.

b. Marton, Claire, *et al. Bibliography of Electron Microscopy.* Washington, D. C.: Government Printing Office, 1950. 87 pp. (U. S. National Bureau of Standards Circular No. 502.)

---

34. E. P. Gray, "A Selective Survey of Books on Plasma Physics." *Physics Today,* 16 (No. 11): 66-74, November 1963.

c. *International Bibliography of Electron Microscopy*, 1950-1961. New York: New York Society of Electron Microscopists, 1959-1962. 2 vols.
This has terminated. (It contains references that were first published by the Society on edge-notch cards quarterly.)

d. Haine, M. E., and Cosslett, Vernon E. *The Electron Microscope; The Present State of the Art.* New York: Interscience Publishers, 1961. 282 pp.

e. Siegel, Benjamin M., editor. *Modern Developments in Electron Microscopy.* New York: Academic Press, 1964. 432 pp.
Physics, methods and applications are covered by eight authorities.

f. Wyckoff, Ralph W. G. *Electron Microscopy; Technique and Applications.* New York: Interscience Publishers, 1949. 248 pp.

g. Burton, E. F., and Kohl, W. H. *The Electron Microscope.* (Second Edition.) New York: Reinhold Publishing Corporation, 1946. 325 pp.
This well-illustrated presentation compares ordinary light microscopy with the electron type.

h. Hall, Cecil E. *Introduction to Electron Microscopy.* (Second Edition.) New York: McGraw-Hill Book Company, 1966. 397 pp.

Electronic circuits.

Tube theory of operation (rather than external circuits) is discussed from the physicist's standpoint in:

Koller, L. R. *The Physics of Electron Tubes.* (Second Edition.) New York: McGraw-Hill Book Company, 1937. 234 pp.

More recent books on tubes and circuits follow:

a. Millman, Jacob, and Halkias, Christos C. *Electronic Devices and Circuits.* New York: McGraw-Hill Book Company, 1967. 752 pp.

b. Kloeffler, Royce G. *Electron Tubes.* New York: John Wiley and Sons, 1966. 262 pp.

c. Angelo, E. J., Jr. *Electronic Circuits; A Unified Treatment of Vacuum Tubes and Transistors.* (Second Edition.) New York: McGraw-Hill Book Company, 1964. 652 pp.

d. Seely, Samuel. *Electronic Circuits.* New York: Holt, Rinehart and Winston, 1968. 752 pp.

See also the section Circuits in Chapter X—Electricity and Magnetism.

Advances in solid state physics are well applied to electronic devices in:

a. Nussbaum, Allen. *Semiconductor Device Physics.* Englewood Cliffs, N. J.: Prentice-Hall, Inc., 1962. 340 pp.

b. McKelvey, J. P. *Solid-State and Semiconductor Physics.* New York: Harper and Row, 1966. 512 pp.

c. Lindmayer, Joseph, and Wrigley, Charles Y. *Fundamentals of Semiconductor Devices.* Princeton, N. J.: D. Van Nostrand Company, 1965. 486 pp.

Pioneer works on the tiny-transistor concept were:

a. Shockley, William. *Electrons and Holes in Semiconductors.* New York: D. Van Nostrand Company, 1950. 558 pp.

b. Shea, Richard F., editor. *Principles of Transistor Circuits.* New York: John Wiley and Sons, 1953. 535 pp.

There is now a comprehensive handbook:

Hunter, Lloyd P., editor. *Handbook of Semiconductor Electronics.* (Second Edition.) New York: McGraw-Hill Book Company, 1962. Various paging.

The Semiconductor Electronics Education Committee, abbreviated "SEEC" and centered at Massachusetts Institute of Technology, issued an interesting series of seven volumes, the first and last being:

a. Adler, R. B.; Smith, A. C.; and Longini, R. L. *Introduction to Semiconductor Physics.* New York: John Wiley and Sons, 1964. 247 pp.

b. Thornton, R. D., *et al. Handbook of Basic Transistor Circuits and Measurements.* New York: John Wiley and Sons, 1966. 156 pp. See its flyleaf for listing of other SEEC volumes.

Among many current books on transistors, etc.:

a. Ristenbatt, Marlin P., and Riddle, Robert L. *Transistor Physics and Circuits.* (Second Edition.) Englewood Cliffs, N. J.: Prentice-Hall, Inc., 1966. 549 pp.

b. Moll, John L. *Physics of Semiconductors.* New York: McGraw-Hill Book Company, 1964. 293 pp.

c. Le Croissette, Dennis. *Transistors.* Englewood Cliffs, N. J.: Prentice-Hall, Inc., 1963. 280 pp.

d. Hunter, Lloyd P. *Introduction to Semiconductor Phenomena and Devices.* Reading, Mass.: Addison-Wesley Publishing Company, 1966. 218 pp.

e. Valdes, Leopoldo B. *The Physical Theory of Transistors.* New York: McGraw-Hill Book Company, 1961. 370 pp.

See also "Resource Letter Scr-1 on Semiconductors," by Paul Handler. *American Journal of Physics,* 32: 329-333, May 1964.

Electronic noise associated with vacuum tubes and semiconductors is discussed in:

a. Van der Ziel, Aldert. *Noise.* Englewood Cliffs, N. J.: Prentice-Hall, Inc., 1954. 450 pp.

b. Bennett, William R. *Electrical Noise.* New York: McGraw-Hill Book Company, 1960. 280 pp.

Electron emission.

Under suitable conditions, electrons will be emitted or dislodged by heat, light, or ion-bombardment.

Stimulated emission yields the spectacular effects of lasers and masers, which are atomic amplifiers or oscillators. "Laser" is derived from initial letters of Light Amplification by Stimulated Emission of Radiation; "Maser" from Microwave Amplification . . . Introductory treatments include:

a. Brotherton, Manfred. *Masers and Lasers; How They Work, What They Do.* New York: McGraw-Hill Book Company, 1964. 207 pp.
A simplified explanation for engineers and scientists.

b. Troup, Gordon. *Masers and Lasers; Molecular Amplification and Oscillation by Stimulated Emission.* (Second Edition.) London: Methuen and Company, 1963. 200 pp. (Methuen Monograph.)

c. Thorp, J. S. *Masers and Lasers: Physics and Design.* New York: St. Martin's Press, 1967. 312 pp.

d. Lengyel, Bela A. *Introduction to Laser Physics.* New York: John Wiley and Sons, 1966. 309 pp.

e. Smith, William V., and Sorokin, Peter P. *The Laser.* New York: McGraw-Hill Book Company, 1966. 498 pp.

For reprints of 450 papers with commentary, see Joseph Weber's two volumes, *Lasers* and *Masers*, published by Gordon and Breach in 1966.

See also "Resource Letter MOP-1 on Masers (Microwave Through Optical) and on Optical Pumping," by H. W. Moos. *American Journal of Physics*, 32: 589-595, August 1964.

Light-electricity conversion of many kinds is discussed in:

Larach, Simon, editor. *Photoelectronic Materials and Devices.* Princeton, N. J.: D. Van Nostrand Company, 1965. 448 pp.
Includes photoconductivity, photoelectric emission, electroluminescence, photovoltaic effect, optoelectronics, etc.

Photoelectric emission is the liberation of electrons from light-sensitive surfaces, as described in:

a. Zworykin, V. K., and Ramberg, E. G. *Photoelectricity and Its Application.* New York: John Wiley and Sons, 1949. 494 pp.

b. Hughes, Arthur L., and DuBridge, Lee A. *Photoelectric Phenomena.* New York: McGraw-Hill Book Company, 1932. 531 pp.

c. Summer, W. *Photosensitors; A Treatise on Photo-Electric Devices and Their Application to Industry.* London: Chapman and Hall, Ltd., 1957. 675 pp.

Several bibliographies bearing on this subject are useful:

a. Doty, Marion F., compiler. *Selenium; A List of References, 1817-1925.* New York: The New York Public Library, 1927. 114 pp.

b. Bell Telephone Laboratories. *Bibliography of Articles on Photoelectricity, 1896-June 1930.* New York: Bell Telephone Laboratories, 1930. 145 pp.

c. Bell Telephone Laboratories. *Photoelectric Cells; Applications, 1913-1942.* New York: Bell Telephone Laboratories, 1942. 229 pp.

Solid-state photoconductivity is described in:

a. Rose, Albert. *Concepts in Photoconductivity and Allied Problems.* New York: Interscience Publishers, 1963. 160 pp. Physical ideas underlying photoconductors.

b. Bube, Richard H. *Photoconductivity of Solids.* New York: John Wiley and Sons, 1960. 461 pp.

Thermal emission, whereby charged particles leave heated filaments within radio and other types of vacuum tube, is surveyed by:

a. Reimann, Arnold L. *Thermionic Emission.* New York: John Wiley and Sons, 1934. 324 pp.

b. Richardson, Owen W. *The Emission of Electricity from Hot Bodies.* (Second Edition.) London: Longmans, Green and Company, 1921. 320 pp.

See also index under Thermoelectricity.

Cathode rays are electron streams emanating from a negative electrode subjected to ion bombardment in a gaseous discharge tube. An introductory description of such tubes is provided by:

Parr, Geoffrey, and Davie, O. H. *The Cathode Ray Tube and Its Applications.* (Third Edition.) New York: Reinhold Publishing Corporation, 1959. 433 pp.

Oscillographs chart electrical relationships by means of charged plates that deflect a cathode beam in perpendicular direction simultaneously. Several treatments are:

a. Rider, John F., and Uslan, Seymour D. *Encyclopedia on*

Cathode-ray Oscilloscopes and Their Uses. (Second Edition.) New York: John F. Rider Publisher, Inc., 1959. Various paging.
This carefully compiled and comprehensive reference work covers theoretical considerations, equipment, circuits, wave-forms, and applications.

b. Wilson, William. *The Cathode Ray Oscillograph in Industry.* (Fourth Edition.) London: Chapman and Hall, Ltd., 1953. 273 pp.

c. Czech, J. *The Cathode Ray Oscilloscope: Circuitry and Practical Applications.* New York: Interscience Publishers, 1957. 340 pp.

d. Ruiter, Jacob H., Jr. *Modern Oscilloscopes and Their Uses.* (Revised Edition.) New York: Rinehart and Company, 1955. 346 pp.

X-rays.

These are high frequency waves emitted from substances hit by cathode rays. An old but still useful bibliography is:

Gocht, Hermann. *Die Röntgen-Literatur.* Stuttgart: Ferdinand Enke, 1911-1921. 4 vols.
This set covers physical and medical aspects, and has author, subject and patent approaches. The literature is surveyed through 1917.

There is also a comprehensive handbook of historic interest:

Marx, Erich. *Handbuch der Radiologie.* Leipzig: Akademische Verlagsgesellschaft, 1920-1925. 6 vols. (A second edition of Vol. 6 has appeared in two parts, 1933-1934, under the title, *Quantenmechanik der Materie und Strahlung.*)

More recent general compendia are:

a. Kaelble, Emmett F., editor. *Handbook of X-Rays, for Diffraction, Emission, Absorption and Microscopy.* New York: McGraw-Hill Book Company, 1967. Var. paged.

b. Clark, George L., editor. *Encyclopedia of X-Rays and Gamma Rays.* New York: Reinhold Publishing Corporation, 1963. 1149 pp.

Theory and practice are presented in:

a. Compton, Arthur H., and Allison, Samuel K. *X-rays in Theory and Experiment.* (Second Edition.) New York: D. Van Nostrand Company, 1935. 828 pp.
Classical and quantum interpretations meet in this overall review.

b. Robertson, John K. *Radiology Physics.* (Second Edition.) New York: D. Van Nostrand Company, 1948. 323 pp.

c. Clark, George L. *Applied X-rays.* (Fourth Edition.) New York: McGraw-Hill Book Company, 1955. 843 pp.

d. St. John, Ancel, and Isenburger, Herbert R. *Industrial Radiology.* (Second Edition.) London: Chapman and Hall, Ltd., 1943. 298 pp.
A noteworthy feature is the comprehensive bibliography (pp. 233-289) of 1314 references, continued by several supplements published separately later.

e. Sproull, Wayne T. *X-rays in Practice.* New York: McGraw-Hill Book Company, 1946. 615 pp.

Special phases of X-ray work are covered by:

a. Cosslett, Vernon E., and Nixon, W. C. *X-Ray Microscopy.* Cambridge, England: At the University Press, 1960. 406 pp.

b. Siegbahn, Manne. *Spektroskopie der Röntgenstrahlen.* (2. Auflage.) Berlin: Springer, 1931. 575 pp.
This is an excellent monograph on experimental X-ray spectroscopy and its significance for atomic physics. On pp. 488-560 literature is cited by years through 1931. (An English translation of the first German edition, 1923, had appeared in 1925.)

c. Laue, Max von, and Wagner, E. H. *Röntgenstrahlinterferenzen.* (Third Edition.) Frankfurt am Main: Akademische Verlagsgesellschaft, 1960. 476 pp.

Diffraction of X-rays and electrons may be effected not only by optical gratings, but also by crystals. This phenomenon has yielded a two-fold contribution to our knowledge of X-rays and crystal structure. Treatments include:

a. Nuffield, E. W. *X-Ray Diffraction Methods.* New York: John Wiley and Sons, 1966. 409 pp.

b. Thomson, Sir George P., and Cochran, W. *Theory and Practice of Electron Diffraction.* London: The Macmillan Company, 1939. 334 pp.

c. Bragg, Sir William H.; Bragg, Sir (William) Lawrence; *et al. The Crystalline State.* London: G. Bell and Sons, 1933-1966. 4 vols., as follows:

Vol. 1: *A General Survey,* by Sir (William) Lawrence Bragg. 1933. 352 pp.

Vol. 2: *The Optical Principles of the Diffraction of X-rays,* by R. W. James. 1948. 623 pp.

Vol. 3: *The Determination of Crystal Structures,* by H. S. Lipson and W. Cochran. (Third Edition.) 1966. 414 pp.

Vol. 4: *Crystal Structure of Minerals,* by Sir (William) Lawrence Bragg and G. F. Claringbull. 1965. 409 pp.
The distinguished Braggs (father and son) jointly received the Nobel

Prize in Physics in 1915 for their work in x-ray spectroscopy and crystal structure.

d. Peiser, H. S.; Rooksby, H. P.; and Wilson, A. J. C., editors. *X-Ray Diffraction by Polycrystalline Materials.* London: Institute of Physics, 1955. 725 pp.

e. Bacon, G. E. *X-Ray and Neutron Diffraction.* Oxford, New York, etc.: Pergamon Press, 1967. 368 pp.

f. Hosemann, R., and Bagchi, S. N. *Direct Analysis of Diffraction by Matter.* New York: Interscience Publishers, 1962. 734 pp.

Basic tables have been revised:

International Union of Crystallography. *International Tables for X-Ray Crystallography.* Birmingham, England: Kynock Press. 3 vols., as follows: Vol. 1: Symmetry groups, 1965. (Second Edition); Vol. 2: Mathematical tables, 1959; and Vol. 3: Physical and chemical tables, 1962.

Originally *Internationale Tabellen zur Bestimmung von Kristallstrukturen.*

See also index entries under Diffraction; Crystallography; Spectroscopy; and Scattering.

Applied electronics.

As previously indicated, this *Guide* cannot deal extensively with applications, such as in radio, television, control, radar, etc. Volumes of the Massachusetts Institute of Technology *Radiation Laboratory Series* [35] are helpful. Other useful sources include:

a. Alley, Charles L., and Atwood, Kenneth W. *Electronic Engineering.* (Second Edition.) New York: John Wiley and Sons, 1966. 760 pp.

b. Terman, Frederick E., *et al. Electronic and Radio Engineering.* (Fourth Edition.) New York: McGraw-Hill Book Company, 1955. 1078 pp.

c. Henney, Keith, editor. *Radio Engineering Handbook.* (Fifth Edition.) New York: McGraw-Hill Book Company, 1959. 1793 pp.

d. Chute, George M. *Electronics in Industry.* (Third Edition.) New York: McGraw-Hill Book Company, 1964. 630 pp.

e. Kiver, Milton S. *Color Television Fundamentals.* (Second Edition.) New York: McGraw-Hill Book Company, 1964. 335 pp.

---

35. Published by McGraw-Hill Book Company, 1947-1953, in 28 vols., the last of which is the index, prepared by Keith Henney.

f. Berkowitz, Raymond S. *Modern Radar: Analysis, Evaluation, and System Design.* New York: John Wiley and Sons, 1966. 660 pp.

Descriptions and circuits of radio and television receivers may be found under model number in the indexes to the Howard W. Sams manuals and to the discontinued John F. Rider series.

Loose-leaf tube and transistor manuals are issued by RCA, GE, and other manufacturers.

8—Molecular, Atomic and Nuclear Physics

General.

Before consulting the standard textbooks, the beginner might well read more popular accounts, such as:

a. Hecht, Selig. *Explaining the Atom.* (Revised Edition.) Enlarged by Eugene Rabinowitch. New York: Viking Press, 1954. 237 pp.

b. Gamow, George. *The Atom and Its Nucleus.* Englewood Cliffs, N. J.: Prentice-Hall, Inc., 1961. 153 pp.

c. Thomson, Sir George P. *The Atom.* (Sixth Edition.) London: Oxford University Press, 1962. 228 pp.

Lucid classics like the following may next be consulted:

a. Millikan, Robert A. *Electrons (+ and −), Protons, Photons, Neutrons, Mesotrons, and Cosmic Rays.* (Fourth Edition.) Chicago: University of Chicago Press, 1947. 642 pp.

b. Fermi, Enrico. *Elementary Particles.* New Haven, Conn.: Yale University Press, 1951. (Reprinted 1961.) 110 pp.

c. Stranathan, J. D. *The "Particles" of Modern Physics.* Philadelphia: The Blakiston Company, 1942. 571 pp.
The "particles" are: electrons; positive rays; photons; positrons; neutrons; mesotrons; X-rays; alpha, beta and gamma rays; cosmic rays.

d. Finkelnburg, Wolfgang. *Structure of Matter,* translated from the 9th/10th German edition. New York: Academic Press, 1964. 511 pp.

e. Crowther, James A. *Ions, Electrons and Ionizing Radiations.* (Eighth Edition.) London: Edward Arnold and Company, 1949. 322 pp.

(We are here chiefly concerned with electrons in atomic structure rather than with their flow discussed in Chapter X—Electronics.)

Recent general introductions include:

a. Frisch, David H., and Thorndike, Alan M. *Elementary Particles.* Princeton, N. J.: D. Van Nostrand Company, 1964. 168 pp.
This was the first in the Momentum Book series sponsored by the Commission on College Physics.

b. Swartz, Clifford E. *The Fundamental Particles.* Reading, Mass.: Addison-Wesley Publishing Company, 1965. 152 pp.

c. Gasiorowicz, Stephen. *Elementary Particle Physics.* New York: John Wiley and Sons, 1966. 613 pp.

d. Williams, W. S. C. *An Introduction to Elementary Particles.* New York: Academic Press, 1961. 406 pp.

See also "Resource Letter SAP-1 on Subatomic Particles," by Clifford E. Swartz. *American Journal of Physics,* 34: 1079-1086, December 1966.

Particles are revealed photographically in:

a. Martin, Charles-Noël. *The Thirteen Steps to the Atom; A Photographic Exploration.* London: George G. Harrap and Company, 1959. 256 pp.

b. Powell, C. F.; Fowler, P. H.; and Perkins, D. H. *The Study of Elementary Particles by the Photographic Method; An Account of the Principal Techniques and Discoveries.* Oxford, New York, etc.: Pergamon Press, 1959. 669 pp.
The atlas of photomicrographs includes many of historical interest.

c. Gentner, W., *et al. An Atlas of Typical Expansion Chamber Photographs.* Oxford, New York, etc.: Pergamon Press, 1954. 199 pp.

See also Rochester and Wilson's book under Cosmic rays.

From among many, the following general textbooks on atomic and nuclear physics may now be chosen:

a. Born, Max. *Atomic Physics.* (Seventh Edition.) New York: Hafner Publishing Co., 1962. 459 pp.

b. Kaplan, Irving. *Nuclear Physics.* (Second Edition.) Reading, Mass.: Addison-Wesley Publishing Company, 1963. 770 pp.

c. Oldenberg, Otto, and Holladay, Wendell G. *Introduction to Atomic and Nuclear Physics.* (Fourth Edition.) New York: McGraw-Hill Book Company, 1967. 414 pp.

d. Rusk, Rogers D. *Introduction to Atomic and Nuclear Physics.* (Second Edition.) New York: Appleton-Century-Crofts, Inc., 1964. 470 pp.

e. Semat, Henry. *Introduction to Atomic and Nuclear Physics.*

(Fourth Edition.) New York: Holt, Rinehart and Winston, 1962. 628 pp.

f. Roy, R. R., and Nigam, B. P. *Nuclear Physics: Theory and Experiment.* New York: John Wiley and Sons, 1967. 616 pp.

g. White, Harvey E. *Introduction to Atomic and Nuclear Physics.* Princeton, N. J.: D. Van Nostrand Company, 1964. 560 pp.

See also "Resource Letter NS-1 on Nuclear Structure," by M. A. Preston. *American Journal of Physics,* 32: 820-824, November 1964.

See also "Resource Letter NR-1 on Nuclear Reactions," by T. A. Griffy. *American Journal of Physics,* 35: 297-301, April 1967.

Collections of published accounts of major hypotheses and discoveries comprise an interesting and unusual compilation:

Foundations of Nuclear Physics; Facsimiles of Thirteen Fundamental Studies as They Were Originally Reported in the Scientific Journals. With a bibliography compiled by Robert T. Beyer. New York: Dover Publications, 1949. 272 pp.

The criteria of selection are frequency of subsequent citation, and amount of research stimulated. The bibliography is arranged in twelve sections, e.g., Radioactivity; Neutrons; Nuclear Fission.

Among molecular phenomena of interest to physicists are capillarity, surface tension, adsorption, cohesion, etc., treated in:

a. Adam, Neil K. *The Physics and Chemistry of Surfaces.* (Third Edition.) London: Oxford University Press, 1941. 436 pp.

b. Semenchenko, V. K. *Surface Phenomena in Metals and Alloys.* Reading, Mass.: Addison-Wesley Publishing Company, 1962. 466 pp.

c. Shewmon, Paul G. *Diffusion in Solids.* New York: McGraw-Hill Book Company, 1963. 203 pp.

d. Ross, Sydney, and Olivier, James P. *On Physical Adsorption.* New York: John Wiley and Sons, 1964. 401 pp.

e. Defay, R.; Progogine, I.; and Bellemans, A. *Surface Tension and Adsorption.* New York: John Wiley and Sons, 1966. 432 pp.

f. Jaswon, M. A. *The Theory of Cohesion: An Outline of the Cohesive Properties of Electrons in Atoms, Molecules, and Crystals.* New York: Interscience Publishers, 1954. 245 pp.

Many-body problem aspects have been treated extensively, as Feshbach states:

> If we include field theory, as in fact we should, it is not an exaggeration to say that theoretical physics in recent years has been almost completely devoted to the study of the ground state and low-lying states of many-body systems; states which empirically are found to have particularly simple properties. The ground states of atomic nuclei, the behavior of electrons in metals, superconductivity, and liquid helium are familiar

examples. It is the aim of theory to predict properties of these states starting from the known elementary forces between the constituent particles.[36]

Related books are:

a. Haar, Dirk ter. *Introduction to the Physics of Many-Body Systems.* New York: Interscience Publishers, 1958. 127 pp.

b. Caianiello, E. R., editor. *Lectures on the Many-Body Problem.* New York: Academic Press, 1962-1964. 2 vols.

Solids have also moved to the forefront of physics interest. In the table of contents of *Solid State Abstracts Journal* we find solid-state physics divided into elastic, acoustic, thermal, optical, dielectric, conductive, magnetic and other properties, which parallel the conventional physics outline and subsume all its familiar topics (superconductivity, luminescence, plasmas, elasticity, ultrasonics, etc.). Gray encountered this dilemma of parallelism while organizing his handbook:

> Adding a chapter so named [Solid-State Physics] to the conventionally labeled group of mechanics, heat, acoustics, and so forth is, of course, a little like trying to divide people into women, men, girls, boys, and zither players. (There was, in fact, one suggestion that perhaps the book should contain only three major sections—Solid-State Physics, Liquid-State Physics, and Gaseous-State Physics.) [37]

Among general introductions are:

a. Kittel, Charles. *Introduction to Solid State Physics.* (Third Edition.) New York: John Wiley and Sons, 1966. 648 pp.

b. Dekker, Adrianus J. *Solid State Physics.* Englewood Cliffs, N. J.: Prentice-Hall, Inc., 1957. 540 pp.

c. Sachs, Mendel. *Solid State Theory.* New York: McGraw-Hill Book Company, 1963. 350 pp.

d. Holden, Alan. *The Nature of Solids.* New York: Columbia University Press, 1965. 241 pp.

e. Seitz, Frederick. *The Modern Theory of Solids.* New York: McGraw-Hill Book Company, 1940. 698 pp.
This is an advanced treatise on "that phase of solid bodies that deals with electronic structure."

See also index under Solid state.

Magnetic-resonance techniques reveal structure:

a. Slichter, C. P. *Principles of Magnetic Resonance.* New York: Harper and Row, 1963. 246 pp.

---

36. H. Feshbach, in review. *Physics Today,* 12 (No. 5): 40, May 1959.
37. D. E. Gray, "The New AIP Handbook." *Physics Today,* 16 (No. 7): 40-42, July 1963.

b. Pake, G. E. *Paramagnetic Resonance.* New York: W. A. Benjamin, Inc., 1962. 205 pp.

c. Al'tshuler, S. A., and Kozyrev, B. M. *Electron Paramagnetic Resonance.* New York: Academic Press, 1964. 369 pp.

d. Abragam, A. *The Principles of Nuclear Magnetism.* Oxford: At the Clarendon Press, 1961. 599 pp.

e. Andrew, E. R. *Nuclear Magnetic Resonance.* Cambridge, England: At the University Press, 1955. 265 pp.

f. Aleksandrov, I. V. *The Theory of Nuclear Magnetic Resonance.* New York: Academic Press, 1966. 197 pp.

g. Emsley, J. W.; Feeney, J.; and Sutcliffe, L. H. *High Resolution Nuclear Magnetic Resonance Spectroscopy.* Oxford, New York, etc.: Pergamon Press, 1965-1966. 2 vols.

h. Hecht, Harry G. *Magnetic Resonance Spectroscopy.* New York: John Wiley and Sons, 1967. 159 pp.

See also "Resource Letter NMR-EPR-1 on Nuclear Magnetic Resonance and Electron Paramagnetic Resonance," by R. E. Norberg. *American Journal of Physics,* 33: 71-75, February 1965.

See also index under Mössbauer effect, and Microwave spectroscopy.

Imperfections and dislocations have positive attributes according to Bailey:

> It would be no exaggeration to say that the factors which make solids useful are the things which are wrong with them. Controlled impurities determine the conductive properties of semiconductors and the hardness of alloys; vacancies and interstituals affect optical and electrical properties and diffusion; surface cracks limit the strength of brittle solids and are important in oxidation and catalysis; line imperfections (dislocations) determine the strength and plastic deformation of metals and permit the growth of solids in the first place.[38]

Relevant books include:

a. Gray, T. J., editor. *The Defect Solid State.* New York: Interscience Publishers, 1957. 511 pp.

b. Weertman, Johannes, and Weertman, Julia R. *Elementary Dislocation Theory.* New York: The Macmillan Company, 1964. 213 pp.

c. Friedel, Jacques. *Dislocations.* Reading, Mass.: Addison-Wesley Publishing Company, 1964. 491 pp.
A comprehensive treatise rather than introduction.

---

38. J. M. Bailey, in review. *American Journal of Physics,* 33: 1091-1092, December 1965.

d. Van Bueren, H. G. *Imperfections in Crystals.* New York: Interscience Publishers, 1960. 676 pp.

See also the index under Metals.

For crystal physics and structure, we have:

a. Zhdanov, G. S. *Crystal Physics,* edited by A. F. Brown. New York: Academic Press, 1966. 500 pp.

b. Buerger, Martin J. *Crystal-Structure Analysis.* New York: John Wiley and Sons, 1960. 668 pp.

c. Wyckhoff, Ralph W. G. *Crystal Structures.* (Second Edition.) New York: Interscience Publishers, 1963-1966. 5 vols.
This supersedes his former loose-leaf compilation. The place of crystallography among the physical sciences is shown in his article, "Crystallography," in *Physics Today,* 5 (No. 10): 4-9, October 1952.

d. Bloss, F. Donald. *An Introduction to the Methods of Optical Crystallography.* New York: Holt, Rinehart and Winston, 1961. 294 pp.

See also John S. Kasper's "Literature of Crystallography," in *Physics Today,* 19 (No. 11): 75-77, November 1966, especially concerning International Union of Crystallography publications, e.g., its tables (previously mentioned), monthly *Acta Crystallographica,* and annual *Structure Reports.*

See also index under Crystallography.

X-ray crystallography has been discussed under the topic diffraction of X-rays.

Atomic energy.

Two convenient compendia are:

a. Glasstone, Samuel. *Sourcebook on Atomic Energy.* (Third Edition.) Princeton, N. J.: D. Van Nostrand Company, 1967. 883 pp.

b. Hogerton, John F. *The Atomic Energy Deskbook.* New York: Reinhold Publishing Corporation, 1963. 673 pp.

Additional help with unfamiliar terminology may be derived from the National Research Council glossary or Del Vecchio's, noted before under Physics Dictionaries in Chapter VIII.

Bibliographical assistance is obtainable from frequently updated USAEC publications such as *What's Available in the Atomic Energy Literature* (TID-4550), and *Bibliographies of Interest . . .* (TID-3043), as well as the following guide:

Anthony, L. J. *Sources of Information on Atomic Energy.* Oxford, London, etc.: Pergamon Press, 1966. 245 pp.
Three shorter editions had emanated from Harwell.

There is also a comprehensive bibliography:

United Nations. Atomic Energy Commission Group. *An International Bibliography on Atomic Energy.* New York: United Nations, 1949-1953. 2 vols., each with 2 supplements.
Volume titles are: 1, Political, economic and social aspects; and 2, Scientific aspects.

High-energy phenomena are described in:

a. Burhop, E. H. S., editor. *High Energy Physics.* New York: Academic Press, 1967. 2 vols.

b. Rossi, Bruno B. *High-Energy Particles.* Englewood Cliffs, N. J.: Prentice-Hall, Inc., 1952. 569 pp.
Cosmic-ray research especially.

Cyclic and linear accelerators (linacs) yield high velocity particles for research on scattering effects, elementary-particle theory, etc.:

a. Livingood, John J. *Principles of Cyclic Particle Accelerators.* Princeton, N. J.: D. Van Nostrand Company, 1961. 392 pp.
Covers the basic theory of cyclic acceleration.

b. Livingston, M. Stanley, and Blewett, John P. *Particle Accelerators.* New York: McGraw-Hill Book Company, 1962. 666 pp.
Supplements Livingood with historical, theoretical and practical material on individual types.

c. Kolomensky, A. A., and Lebedev, A. N. *Theory of Cyclic Accelerators.* New York: Interscience Publishers, 1966. 403 pp.

See also "Resource Letter PA-1 on Particle Accelerators," by John P. Blewett. *American Journal of Physics,* 34: 742-752, September 1966.

Beams (molecular or atomic) are conveyed in beam-transport systems for experimental deflection, scattering, etc. Relevant books include:

a. Steffen, Klaus G. *High Energy Beam Optics.* New York: Interscience Publishers, 1965. 211 pp.
Beam transport and spectrometer design work.

b. Pierce, John R. *Theory and Design of Electron Beams.* (Second Edition.) Princeton, N. J.: D. Van Nostrand Company, 1954. 222 pp.

c. Ramsey, Norman F. *Molecular Beams.* Oxford: At the Clarendon Press, 1956. 466 pp.

d. Banford, Anthony P. *The Transport of Charged Particle Beams.* London: E. and F. N. Spon, 1966. 229 pp.

e. Septier, Albert. *Focusing of Charged Particles.* New York: Academic Press, 1967. 2 vols.
For particle-accelerator and electron-optics work.

f. Bakish, Robert, editor. *Introduction to Electron Beam Technology.* New York: John Wiley and Sons, 1962. 452 pp.

See also "Resource Letter MB-1 on Experiments with Molecular Beams," by Jens C. Zorn. *American Journal of Physics,* 32: 721-732, October 1964.

For scattering and collision phenomena see:

a. Newton, Roger G. *Scattering Theory of Waves and Particles.* New York: McGraw-Hill Book Company, 1966. 681 pp.
A unified treatment of electromagnetic, classical particle, and quantum particle scattering.

b. Rodberg, Leonard S., and Thaler, Roy M. *Introduction to the Quantum Theory of Scattering.* New York: Academic Press, 1967. 398 pp.

c. Massey, H. S. W., and Burhop, E. H. S. *Electronic and Ionic Impact Phenomena.* London: Oxford University Press, 1952. 669 pp.
Classic on charged-particle collision phenomena. (Burhop's *High Energy Physics* was mentioned above.)

d. Mott, Sir Nevill F., and Massey, H. S. W. *The Theory of Atomic Collisions.* (Third Edition.) London: Oxford University Press, 1965. 858 pp.
Advanced presentation for graduate students.

e. McDaniel, Earl W. *Collision Phenomena in Ionized Gases.* New York: John Wiley and Sons, 1964. 775 pp.

f. Goldberger, Marvin L., and Watson, Kenneth M. *Collision Theory.* New York: John Wiley and Sons, 1964. 919 pp.
Scattering fully covered.

g. Hasted, J. B. *Physics of Atomic Collisions.* London: Butterworths Scientific Publications, 1964. 536 pp.

h. Bates, D. R., editor. *Atomic and Molecular Processes.* New York: Academic Press, 1962. 904 pp.
Radiative and collision phenomena are thoroughly treated by twenty-one investigators.

i. Herman, Robert, and Hofstadter, Robert. *High-Energy Electron Scattering Tables.* Stanford, Cal.: Stanford University Press, 1960. 278 pp.

See also index under Kinetic theory, and Scattering, light.

Chain-reacting piles yield nuclear energy, but also isotopes and high-speed neutrons for research, as summarized in:

a. Grant, P. J. *Elementary Reactor Physics.* New York: Pergamon Press, 1966. 196 pp.

b. Glasstone, Samuel, and Edlund, M. C. *The Elements of Nuclear Reactor Theory.* New York: D. Van Nostrand Company, 1952. 416 pp.

c. Marion, J. B., and Fowler, J. L. *Fast Neutron Physics.* New York: Interscience Publishers, 1960-1963. 2 vols.
Part 1: Techniques (1007 pp.); Part 2: Experiments and Theory (1308 pp.).

d. Jakeman, D. *Physics of Nuclear Reactors.* New York: American Elsevier Publishing Company, 1966. 356 pp.

e. Lamarsh, John R. *An Introduction to Nuclear Reactor Theory.* Reading, Mass.: Addison-Wesley Publishing Company, 1966. 585 pp.

f. Beck, Clifford K. *Nuclear Reactors for Research.* Princeton, N. J.: D. Van Nostrand Company, 1957. 267 pp.

See also "Discovery of Fission" by O. R. Frisch and J. A. Wheeler in *Physics Today,* 20 (No. 11): 43-52, November 1967; the early review by H. A. Bethe, *et al.,* in *Reviews of Modern Physics* for April 1936 and April-July 1937; R. T. Beyer's *Foundations of Nuclear Physics;* experimental-nucleonics books in Chapter V; and *Nuclear Science Abstracts,* current indexing medium.

Radiation [39] and other hazards are considered in:

a. Hollaender, Alexander, editor. *Radiation Biology.* New York: McGraw-Hill Book Company, 1954-1956. 3 vols.
Volume contents: 1, Ionizing radiations; 2, Ultraviolet radiations; and 3, Visible and near-visible light. Monumental reference set that discusses radiation effects from physical, chemical and biological viewpoints.

b. Rees, D. J. *Health Physics: Principles of Radiation Detection.* Cambridge, Mass.: The M.I.T. Press, 1967. 242 pp.

c. *Safe Handling of Radioactive Materials.* Washington, D. C.: Government Printing Office, 1964. 107 pp. (U. S. National Bureau of Standards Handbook No. 92.)

d. Goldstein, Herbert. *Fundamental Aspects of Reactor Shielding.* Reading, Mass.: Addison-Wesley Publishing Company, 1959. 416 pp.

---

39. An interesting résumé of common exposures is furnished by: F. P. Cowan, "Everyday Radiation." *Physics Today,* 5 (No. 10): 10-16, October 1952.

e. McCullough, C. R. *Safety Aspects of Nuclear Reactors.* Princeton, N. J.: D. Van Nostrand Company, 1957. 237 pp.
Treats radiation within and outside reactor plants; development of safety criteria; reactor accidents and their consequencies; etc.

f. Thompson, T. J., and Beckerley, J. G., editors. *The Technology of Nuclear Reactor Safety.* Vol. 1: *Reactor Physics and Control.* Cambridge, Mass.: The M.I.T. Press, 1964. 743 pp.

The U. S. Atomic Energy Commission revises from time to time its reference list, *Reactor Safety—A Literature Search,* No. TID-3525. The International Atomic Energy Agency also issues manuals, such as *Safe Operation of Critical Assemblies and Research Reactors.*

See also index under Radiation, and the following section.

Radioactivity.

Uranium and other heavy elements are constantly emitting alpha, beta and gamma rays as they slowly disintegrate. This process is called natural radioactivity, as distinct from artificial radioactivity in which unstable nuclei produced by neutron bombardment behave similarly. Pioneer work by the Curies is described in:

Curie, Marie Sklodowska. *Radioactivité.* Paris: Hermann et Cie., 1935. 563 pp.

Recent introductory treatments include:

a. Hermias, Mary, and Joecile, Mary. *Radioactivity; Fundamentals and Experiments.* New York: Holt, Rinehart and Winston, 1963. 209 pp.

b. Mann, Wilfrid B., and Garfinkel, S. B. *Radioactivity and Its Measurement.* Princeton, N. J.: D. Van Nostrand Company, 1966. 168 pp. (Momentum Book.)

Comprehensive classics distinguished by clarity of presentation are:

a. Rutherford, Ernest R.; Chadwick, James; and Ellis, C. D. *Radiations from Radioactive Substances.* Reprinted with corrections. Cambridge, England: At the University Press, 1951. 588 pp.

b. Hevesy, George, and Paneth, F. A. *A Manual of Radioactivity.* (Second Edition.) London: Oxford University Press, 1938. 306 pp.

c. Moon, Philip B. *Artificial Radioactivity.* Cambridge, England: At the University Press, 1949. 102 pp.

See also index under Radioactivity.

Isotopes.

Most elements are comprised of isotopes, which have different atomic weights but the same atomic number in the periodic table.

(Atomic numbers represent number of protons in the nucleus.) Isotope mass measurements are outlined in:

a. Aston, Francis W. *Mass Spectra and Isotopes.* (Second Edition.) New York: Longmans, Green and Company, 1942. 276 pp.

b. *Mass Spectoscopy in Physics Research.* Washington, D. C.: Government Printing Office, 1953. 273 pp. (U. S. National Bureau of Standards Circular No. 522.)

c. Duckworth, Henry E. *Mass Spectroscopy.* Cambridge, England: At the University Press, 1958. 206 pp.

d. Jayaram, R. *Mass Spectrometry; Theory and Application.* New York: Plenum Press, 1966. 225 pp.

For mass spectrometry literature see *Index and Bibliography of Mass Spectrometry, 1963-1965,* compiled by F. W. McLafferty and J. Pinzelick. (Interscience, 1967.)

There is a guide to nuclear compilations:

Gibbs, Roswell C., and Way, Katherine. *A Directory to Nuclear Data Tabulations.* Washington, D. C.: Government Printing Office, 1958. 185 pp.

Supplements may be found in the 1959 and 1960 *Nuclear Data Tables,* mentioned below.

A chronological series of nuclear data follows:

a. *Nuclear Data.* Washington, D. C.: Government Printing Office, 1950. 309 pp. (U. S. National Bureau of Standards Circular No. 499.) Also *Supplements* 1-3, 1950-1951.

b. New Nuclear Data Cumulations appended to *Nuclear Science Abstracts* during the period 1952-1957. *Nuclear Data Tables* for 1959 and 1960 later appeared as separate USAEC publications.

c. *Nuclear Data Sheets* issued 1958-1965 by the NRC Nuclear Data Project, vols. 1-6. These have been reprinted by Academic Press to match continuance in its new journal *Nuclear Data,* Section B, beginning 1966.

Other charts [40] and tables useful in nuclear physics are:

a. Sullivan, William H. *Trilinear Chart of Nuclides.* (Second Edition.) Washington, D. C.: Government Printing Office, 1957. 4 pp. The sheets are mounted in accordion style for convenient use.

b. Hughes, Donald J., and Schwartz, Robert B. *Neutron Cross Sections.* (Second Edition.) Washington, D. C.: Government Printing Office, 1958. 373 pp.

---

40. See also General Electric's frequently updated wall-chart of the nuclides (7th ed.: 1964).

Supplements were published, 1960 and 1965. See also Hughes' explanation of cross-section theory in his 182-page book of same title (Pergamon, 1957).

c. Kunz, Wunibald, and Schintlmeister, Josef. *Nuclear Tables.* Oxford, New York, etc.: Pergamon Press, 1963-1967. 2 parts in 4 vols., as follows:

Part I: *Nuclear Properties,* Vols. 1-2; Part II: *Nuclear Reactions,* Vols. 1-2. (Arranged by elements.)

d. Hyde, Earl K.; Perlman, Isadore; and Seaborg, Glenn T. *The Nuclear Properties of the Heavy Elements.* Englewood Cliffs, N. J.: Prentice-Hall, Inc., 1964. 3 vols.

Volume titles: 1, Systematics of nuclear structure and radioactivity; 2, Detailed radioactivity properties; and 3, Fission phenomena.

e. International Atomic Energy Agency. *International Directory of Isotopes.* (Third Edition.) Vienna: The Agency, 1964. 487 pp.

f. Lederer, Charles M.; Hollander, Jack M.; and Perlman, Isadore. *Table of Isotopes.* (Sixth Edition.) New York: John Wiley and Sons, 1967. 594 pp.

Cosmic rays.

These are charged particles radiated towards earth from outer space. Their physics is described in:

a. Rossi, Bruno B. *Cosmic Rays.* New York: McGraw-Hill Book Company, 1964. 268 pp.

Authoritative introduction by the discoverer of high-energy particles in cosmic rays. For extensive research see his *High-Energy Particles.*

b. Cranshaw, T. E. *Cosmic Rays.* London: Oxford University Press, 1963. 126 pp.

c. Sandström, A. E. *Cosmic Ray Physics.* New York: Interscience Publishers, 1965. 421 pp.

d. Wolfendale, A. W. *Cosmic Rays.* New York: Philosophical Library, 1963. 222 pp.

e. Hopper, V. D. *Cosmic Radiation and High Energy Interactions.* Englewood Cliffs, N. J.: Prentice-Hall, Inc., 1964. 234 pp.

See also "Resource Letter CR-1 on Cosmic Rays," by J. R. Winckler and D. J. Hofmann. *American Journal of Physics,* 35: 2-12, January 1967.

Superbly reproduced pictures of cloud chamber radiation are found in:

Rochester, G. D., and Wilson, J. G.[41] *Cloud Chamber Photographs of the Cosmic Radiation.* New York: Academic Press, 1952. 128 pp.

See also W. Gentner's *Atlas . . .*

Applied atomics.

Military, industrial and medical potentialities of atomic energy are great. The first comprehensive report on the atomic bomb was issued as a U. S. government publication, and also by the Princeton University press:

Smyth, Henry De Wolf. *Atomic Energy for Military Purposes.* Princeton, N. J.: Princeton University Press, 1945. 264 pp.

The early history of nuclear energy is superbly related in:

Hewlett, Richard G., and Anderson, Oscar E., Jr. *History of the United States Atomic Energy Commission.* Vol. 1: *The New World, 1939/1946.* University Park, Pa.: The Pennsylvania State University Press, 1962. 766 pp.

More popular accounts of our atomic development, with biographical sidelights, include:

a. Laurence, William L. *Men and Atoms: The Discovery, the Uses and the Future of Atomic Energy.* New York: Simon and Schuster, Inc., 1962. 319 pp.

b. Compton, Arthur H. *Atomic Quest.* New York and London: Oxford University Press, 1956. 370 pp.

c. Jungk, Robert. *Brighter Than a Thousand Suns: A Personal History of the Atomic Scientists.* New York: Harcourt, Brace and Company, 1958. 383 pp.

British activity is related in this chronological triad:

a. Gowing, Margaret. *Britain and Atomic Energy, 1939-1945.* London: The Macmillan Company, 1964. 464 pp.
The first part of a detailed history by the official U. K. archivist.

b. *Harwell; the British Atomic Energy Research Establishment, 1946-1951.* New York: Philosophical Library, 1952. 128 pp.

c. Jay, K. E. B. *Atomic Energy Research at Harwell.* New York: Philosophical Library, 1955. 144 pp.
This covers work at Harwell from 1951 to 1954.

---

41. See also J. G. Wilson, *The Principles of Cloud-Chamber Technique.* Cambridge, England: At the University Press, 1951. 131 pp.

For the Russian account see:

Kramish, Arnold. *Atomic Energy in the Soviet Union.* Stanford, Cal.: Stanford University Press, 1959. 232 pp.

Engineering merges with physics in:

a. Murray, Raymond L. *Introduction to Nuclear Engineering.* (Second Edition.) Englewood Cliffs, N. J.: Prentice-Hall, Inc., 1961. 384 pp.

b. Glasstone, Samuel, and Sesonske, Alexander. *Nuclear Reactor Engineering.* Princeton, N. J.: D. Van Nostrand Company, 1963. 830 pp.

c. Etherington, Harold, editor. *Nuclear Engineering Handbook.* New York: McGraw-Hill Book Company, 1958. 1882 pp.

d. U. S. Atomic Energy Commission. *Reactor Handbook.* (Second Edition.) New York: Interscience Publishers, 1960-1964. 4 vols. in 5.
Volume titles are: 1, Materials; 2, Fuel reprocessing; 3A, Physics; 3B, Shielding; and 4, Engineering.

e. Stephenson, Richard. *Introduction to Nuclear Engineering.* (Second Edition.) New York: McGraw-Hill Book Company, 1958. 491 pp.

Applications and predictions abound in:

a. Crowther, J. G. *Nuclear Energy in Industry.* New York: Pitman Publishing Corporation, 1956. 168 pp.

b. Hughes, Donald J. *On Nuclear Energy; Its Potential for Peacetime Uses.* Cambridge, Mass.: Harvard University Press, 1957. 263 pp.

c. Thomson, Sir George P. *The Foreseeable Future.* Cambridge, England: At the University Press, 1955. 166 pp.

d. Titterton, E. W. *Facing the Atomic Future.* New York: St. Martin's Press, 1956. 378 pp.

e. Wendt, Gerald. *The Prospects of Nuclear Power and Technology.* Princeton, N. J.: D. Van Nostrand Company, 1957. 348 pp.

f. Mann, Martin. *Peacetime Uses of Atomic Energy.* New York: Thomas Y. Crowell Company, 1957. 175 pp.

The peaceful uses of atomic energy are further discussed in *International Conferences* thereon, held in Geneva in 1955, 1958 and 1964 thus far, with proceedings published by United Nations in 17, 33 and 16 vols., respectively. The U. S. Atomic Energy Commission also prepares fine presentation sets on related subjects, the most recent volume-titles being: 1, Research, U.S.A.; 2, Nuclear power, U.S.A.; 3, Radioisotopes and radiation; and 4, Education and the atom; all published by McGraw-Hill.

9—Related Fields

Astrophysics.

This is the region in which astronomy and physics overlap.

a. Brandt, John C., and Hodge, Paul. *Solar System Astrophysics.* New York: McGraw-Hill Book Company, 1964. 448 pp.
Comprehensive introduction to the physics of solar system phenomena.

b. Hynek, J. A., editor. *Astrophysics: A Topical Symposium.* New York: McGraw-Hill Book Company, 1951. 703 pp.

c. McMahon, Allen J. *Astrophysics and Space Science: An Integration of Sciences.* Englewood Cliffs, N. J.: Prentice-Hall, Inc., 1965. 444 pp.

Astrophysical research is no longer earthbound:

a. Liller, William. *Space Astrophysics.* New York: McGraw-Hill Book Company, 1961. 268 pp.
Investigations from above the atmosphere, by means of artificial satellites and space probes.

b. LeGalley, Donald P., and Rosen, Alan, editors. *Space Physics.* New York: John Wiley and Sons, 1964. 752 pp.
Present knowledge of interplanetary space, and research in progress. *Space Science* (1963) has additional U. of California lectures.

c. Lundquist, Charles A. *The Physics and Astronomy of Space Science.* New York: McGraw-Hill Book Company, 1966. 116 pp.
Introductory space science related to satellites, with their applications.

d. Glasstone, Samuel. *Sourcebook on the Space Sciences.* Princeton, N. J.: D. Van Nostrand Company, 1965. 937 pp.

e. Berkner, Lloyd V., and Odishaw, Hugh, editors. *Science in Space.* New York: McGraw-Hill Book Company, 1961. 458 pp.

See also index under Astronautics.

Of historical interest is:

Handbuch der Astrophysik, herausgegeben von G. Eberhard, *et al.* Berlin: Springer, 1928-1936. 7 vols. in 10.

Five volumes (50-54) of Flügge's *Handbuch der Physik* constitute a more recent reference set on astrophysics. There is also a 1965 volume, *Astronomy and Astrophysics,* in Landolt-Börnstein (New Series).

Biophysics.

In a useful overall summary,[42] Loofbourow characterizes this field as follows:

> . . . Biophysics may be said to include all applications of physics to the study or explanation of biological systems. Biophysics so defined may conveniently be divided into three aspects, as follows: (i) the physics of living organisms, (ii) the biological effects of physical agents, and (iii) the use of physical methods and measurements in the study of biological structures and functions.

"Physics and Biology—Where Do They Meet?" is asked by Walter A. Rosenblith in *Physics Today,* 19 (No. 1): 23-34, January 1966.

Among introductory treatments of biophysics are:

a. Epstein, Herman T. *Elementary Biophysics: Selected Topics.* Reading, Mass.: Addison-Wesley Publishing Company, 1963. 122 pp. Topics include biometry; physics of vision, hearing and muscles; and biophysical methods.

b. Setlow, Richard B., and Pollard, Ernest C. *Molecular Biophysics.* Reading, Mass.: Addison-Wesley Publishing Company, 1962. 545 pp.

c. Casey, E. J. *Biophysics: Concepts and Mechanisms.* New York: Reinhold Publishing Corporation, 1962. 352 pp.
Covers bioenergetics, kinetics, ultrasonics, ionizing radiations, macromolecules, etc.

d. Glasser, Otto, editor. *Medical Physics.* Chicago: The Year Book Publishers, 1944-1960. 3 vols.
The third volume supplements and updates the first two. The set has much material of interest in general physics, despite the "medical" title.

See also "Resource Letter PB-1 on Physics and Biology," by D. James Baker, Jr. *American Journal of Physics,* 34: 83-93, February 1966.

Chemical physics.

An introduction to this dual field is:

Slater, J. C. *Introduction to Chemical Physics.* New York: McGraw-Hill Book Company, 1939. 521 pp.
The author explains the union of the two branches as follows (p. v):

> It is probably unfortunate that physics and chemistry ever were separated. Chemistry is the science of atoms and of the way they combine. Physics deals with the interatomic forces and with the large-scale properties of matter resulting from those forces. So long as

---

42. J. R. Loofbourow, "Biophysics." *American Journal of Physics,* 15: 21-30, January-February 1947.

chemistry was largely empirical and nonmathematical, and physics had not learned to treat small-scale atomic forces, the two sciences seemed widely separated. But with statistical mechanics and the kinetic theory on the one hand and physical chemistry on the other, the two sciences began to come together. Now that statistical mechanics has led to quantum theory and wave mechanics, with its explanations of atomic interactions, there is really nothing separating them any more. . . . For want of a better name, since Physical Chemistry is already preempted, we may call this common field Chemical Physics.

Typical subject material may be found in:

Prock, Alfred, and McConkey, Gladys. *Topics in Chemical Physics,* based on the Harvard Lectures of Peter Debye. New York: American Elsevier Publishing Company, 1962. 277 pp.

Note that there is a *Journal of Chemical Physics.*

For further coverage see several chemical literature guides cited in the General Bibliography, appended.

Geophysics.

Various phases of geophysics are spanned by:

Physics of the Earth. Washington, D. C.: National Research Council, 1931-1942. 9 vols.

Individual volumes are: I, Volcanology; II, Figure of the earth; III, Meteorology; IV, Age of the earth; V, Oceanography; VI, Seismology; VII, Internal constitution of the earth; VIII, Terrestrial magnetism and electricity; IX, Hydrology.

The prefaces state:

It is generally agreed that more attention should be given to research in the middle ground between the sciences. Geophysics—the study by physical methods of the planet on which we live—is a conspicuous example of such middle-ground science, as it slides off imperceptibly in one direction or another into the fields of physics, astronomy, and geology, to say nothing of biology, with which the subject of oceanography is closely connected. Some branches of geophysics, such as meteorology, terrestrial magnetism, geodesy, and oceanography, have long been studied more or less independently, but it has become increasingly clear that these subjects and many others are all parts of geophysics.

The preceding set should not be confused with *Physics and Chemistry of the Earth,* a progress series issued by Pergamon Press since 1956.

Handbooks are available:

a. U. S. Air Force. Cambridge Research Laboratories. *Handbook of Geophysics and Space Environments,* edited by Shea L. Valley. New York: McGraw-Hill Book Company, 1966. 683 pp.

b. *Handbook of Physical Constants,* edited by Sydney P. Clark, Jr.

(Revised Edition.) New York: Geological Society of America, 1966. 587 pp.

Current research is described in:

a. Odishaw, Hugh, editor. *Research in Geophysics.* Cambridge, Mass.: The M.I.T. Press, 1964. 2 vols.

b. Runcorn, S. K., editor. *Methods and Techniques in Geophysics.* New York: Interscience Publishers, 1960-1966. 2 vols. (More volumes will appear in this progress series.)

Pergamon Press published *Annals of the International Geophysical Year* (IGY), which was 1957-1958 at peak sunspot activity. (There were other special years, *e.g.*, the International Years of the Quiet Sun (IQSY), 1964-1965.)

Various meteorological compendia are helpful:

a. Berry, F. A., *et al.*, editors. *Handbook of Meteorology.* New York: McGraw-Hill Book Company, 1945. 1116 pp.

b. American Meteorological Society. *Compendium of Meteorology,* edited by Thomas F. Malone. Boston: The Society, 1951. 1334 pp. This is both comprehensive and authoritative.

c. American Meteorological Society. *Glossary of Meteorology,* edited by Ralph E. Huschke. Boston: The Society, 1959. 638 pp.

d. *Smithsonian Meteorological Tables,* edited by R. J. List. (Sixth Edition.) Washington, D. C.: Smithsonian Institution, 1951. 527 pp.

Atmospheric effects are described in:

a. Humphreys, William J. *Physics of the Air.* (Third Edition.) New York: McGraw-Hill Book Company, 1940. 676 pp. Its five parts are: Mechanics and thermodynamics of the atmosphere; Atmospheric electricity and auroras; Meteorological acoustics; Atmospheric optics; Factors of climatic control.

b. Riehl, Herbert. *Introduction to the Atmosphere.* New York: McGraw-Hill Book Company, 1965. 300 pp.

GENERAL SUMMARY

Every special subject guide is expected to point out important landmarks and indicate how to proceed beyond. Two methods of compilation are possible: (1) all book citations not too obviously disqualified might be copied from library card catalogs so as to form an imposing bibliographical array termed "comprehensive"; or (2) books could be carefully examined, compared, selected and organized for the purpose of guiding rather than impressing the user. The latter alternative has been adopted.

Accordingly, the materials included constitute points of departure rather than the whole printed record. This guide attempts to cover most important aspects and topics of physics literature by citing a few representative sources in each category. Library tools and techniques have been described. Attention has constantly been called to special types of publication which fulfill a unique purpose, such as Magie's *Source Book,* Higgins' biographical lists, Sutton's *Demonstration Experiments,* Smith's *Careers in Physics,* etc. These useful and interesting compilations are too frequently overlooked.

As physics literature becomes increasingly voluminous, the importance of knowing how to utilize its hidden resources efficiently grows apace. Library guides in special subject fields have been characterized as signposts. If the present guide beckons users toward the desired ends of their literature searches, it will have achieved its purpose.

GENERAL BIBLIOGRAPHY

The following references have proved helpful in the preparation of this *Guide*, and may be used in conjunction with it:

American Chemical Society. *Searching the Chemical Literature.* (Second Edition.) Washington, D. C.: The Society, 1961. 326 pp.

Bottle, R. T. *Use of the Chemical Literature.* London: Butterworths Scientific Publications, 1962. 231 pp.

Burke, A. J., and Burke, M. A. *Documentation in Education:* Revision of C. Alexander and A. J. Burke: *How to Locate Educational Information and Data.* (Fourth Edition, Revised.) New York: Teachers College Press, 1967. 413 pp.

Burman, C. R. *How to Find Out in Chemistry.* (Second Edition.) Oxford, London, etc.: Pergamon Press, 1966. 226 pp.

Chandler, G. *How to Find Out.* (Third Edition.) Oxford, London, etc.: Pergamon Press, 1967. 198 pp.

Crane, E. J.; Patterson, A. M.; and Marr, E. B. *A Guide to the Literature of Chemistry.* (Second Edition.) New York: John Wiley and Sons, 1957. 397 pp.

Dyson, G. M. *A Short Guide to Chemical Literature.* (Second Edition.) London: Longmans, Green and Company, 1959. 157 pp.

Holmstrom, J. E. *Records and Research in Engineering and Industrial Science.* (Third Edition.) London: Chapman and Hall, Ltd., 1956. 491 pp.

Houghton, B. *Technical Information Sources; A Guide to Patents Standards and Technical Reports Literature.* Hamden, Conn.: Archon Books, 1967. 101 pp.

Jenkins, F. B. *Science Reference Sources.* (Fourth Edition.) Champaign, Illinois: Illini Union Bookstore, 1965. 143 pp.

Johnson, I. *Selected Books and Journals in Science and Engineering.* (Second Edition.) Cambridge, Mass.: The M.I.T. Press, 1959. 63 pp.

Malclès, L.-N. *Les Sources du Travail Bibliographique.* Vol. 3: *Bibliographies Spécialisées (Sciences Exactes et Techniques).* Geneva: Droz, 1958. 575 pp.

Malinowsky, H. R. *Science and Engineering Reference Sources.* Rochester, N. Y.: Libraries Unlimited, 1967. 213 pp.

Mellon, M. G. *Chemical Publications; Their Nature and Use.* (Fourth Edition.) New York: McGraw-Hill Book Company, 1965. 324 pp.

Parke, N. G. *Guide to the Literature of Mathematics and Physics, Including Related Works on Engineering Science.* (Second Edition.) New York: Dover Publications, 1958. 436 pp.

Pemberton, J. E. *How to Find Out in Mathematics.* Oxford, London, etc.: Pergamon Press, 1963. 158 pp.

Roberts, A. D. *Guide to Technical Literature.* London: Grafton and Company, 1939. 279 pp.

Roberts, A. D. *Introduction to Reference Books.* (Third Edition.) London: The Library Association, 1956. 237 pp.

Soule, B. A. *Library Guide for the Chemist.* New York: McGraw-Hill Book Company, 1938. 302 pp.

Spratt, H. P. *Libraries for Scientific Research in Europe and America.* London: Grafton and Company, 1936. 227 pp.

Strauss, L. J., *et al. Scientific and Technical Libraries; Their Organization and Administration.* New York: Interscience Publishers, 1964. 398 pp.

Thornton, J. L., and Tully, R. I. J. *Scientific Books, Libraries and Collectors.* (Second Edition.) London: The Library Association, 1962. 406 pp.

Walford, A. J. *Guide to Reference Material.* (Second Edition.) Vol. 1: *Science and Technology.* London: The Library Association, 1966. 483 pp.

Whitford, R. H., and O'Farrell, J. B. "Use of a Technical Library." *Mechanical Engineering,* 70: 987-993, December 1948.

Winchell, C. M. *Guide to Reference Books.* (Eighth Edition.) Chicago: American Library Association, 1967. 741 pp.

Yates, B. *How to Find Out about Physics.* Oxford, London, etc.: Pergamon Press, 1965. 175 pp.

AUTHOR INDEX

SUBJECT INDEX